ACTA MECHANICA
SUPPLEMENTUM 2

O. E. BARNDORFF-NIELSEN AND B. B. WILLETTS (EDS.)

Aeolian Grain Transport 2
The Erosional Environment

C. Christiansen, J. C. R. Hunt, K. Hutter, J. T. Møller,
K. R. Rasmussen, M. Sørensen, F. Ziegler
(Associate Editors)

SPRINGER-VERLAG WIEN GMBH

Prof. Ole E. Barndorff-Nielsen

Department of Theoretical Statistics, Institute of Mathematics, University of Aarhus,
Aarhus, Denmark

Prof. Brian B. Willetts

Department of Engineering, University of Aberdeen, Kings College, Aberdeen,
United Kingdom

© 1991 by Springer-Verlag Wien
Originally published by Springer-Verlag Wien New York in 1991

With 104 Figures

Library of Congress Cataloging-in-Publication Data. Aeolian grain transport / O. E. Barndorff-Nielsen and B. B. Willetts, eds. ;
C. Christiansen ... [et al.] associate editors. p. cm. — (Acta mechanica. Supplementum, ISSN 0939-7906 ; 1-2). Contents:
1. Mechanics — 2. The erosional environment. ISBN 0-387 82269-0 (v. 1 : N. Y.).
ISBN 978-3-211-82274-6   ISBN 978-3-7091-6703-8 (eBook)
DOI 10.1007/978-3-7091-6703-8
1. Eolian processes. 2. Wind erosion. I. Barndorff-
Nielsen, O. E. (Ole E.) II. Willetts, B. B. (Brian B.) III. Series. GB611.A36 1991. 551.3'7 — dc20. 91-29296

ISSN 0939-7906
ISBN 978-3-211-82274-6

Dedicated to the memory of Paul Robert Owen

Foreword

Wind erosion has such a pervasive influence on environmental and agricultural matters that academic interest in it has been continuous for several decades. However, there has been a tendency for the resulting publications to be scattered widely in the scientific literature and consequently to provide a less coherent resource than might otherwise be hoped for. In particular, cross-reference between the literature on desert and coastal morphology, on the deterioration of wind affected soils, and on the process mechanics of the grain/air-flow system has been disappointing.

A successful workshop on "The Physics of Blown Sand", held in Aarhus in 1985, took a decisive step in collecting a research community with interests spanning geomorphology and grain/wind process mechanics. The identification of that Community was reinforced by the Binghampton Symposium on Aeolian Geomorphology in 1986 and has been fruitful in the development of a number of international collaborations. The objectives of the present workshop, which was supported by a grant from the NATO Scientific Affairs Division, were to take stock of the progress in the five years to 1990 and to extend the scope of the community to include soil deterioration (and dust release) and those beach processes which link with aeolian activity on the coast.

The meeting satisfied these objectives and proved most stimulating. Presentations described both completed studies and work in progress, drawn from many countries and several disciplines. They provided stimulating discussions, the fruit of which will become evident during the next few years. The standard of presentations was generally high and most of them are represented in this collection of papers. Authors have had opportunity to incorporate the outcome of discussion where this was appropriate. So much good material was produced that the papers extend to two volumes.

The subject matter has been structured so that papers appear in cognate groupings because it was felt that this would assist readers, particularly those for whom this is a new field. The two volumes are, accordingly, headed respectively "Mechanics" and "The Erosional Environment", and each contains sub-sections, as can be seen in the table of contents. However, we found difficulty in assigning some sub-sections to a particular volume and some papers to a particular sub-section. This we regard as an encouraging sign that the boundaries between different views, associated with different disciplines and fields of application, are becoming more pervious, and cross-fertilisation more active.

The study of aeolian processes is at a truly exciting stage: the primary applications — desertification, coastal defence, dunefield behaviour — have never been more important. We hope that many people will read these papers, find them interesting and challenging, and will correspond with the authors, to the benefit of the development of understanding of the subject.

During 1990 the deaths occurred of both Ralf Alger Bagnold and Paul Robert Owen, two outstanding contributors to this subject, Bagnold pre-eminently so. Both of them attended the 1985 Workshop, and both would have attended this one had their health permitted. The two deaths greatly diminish the scientific community and, as you will see, we have respectfully dedicated one of these volumes to the memory of each of them.

O. E. Barndorff-Nielsen
B. B. Willetts

January 1991

Contents

Acta Mechanica (1991) [Suppl] 2: 1—22

Air flow and sand transport over sand-dunes

W. S. Weng, J. C. R. Hunt and **D. J. Carruthers**, Cambridge, **A. Warren** and **G. F. S. Wiggs**,
London, **I. Livingstone**, Coventry, and **I. Castro**, Guildford, United Kingdom

Summary. Developments in the modelling of turbulent wind over hills and sand dunes of different
shapes by Hunt et al. [1], Carruthers et al. [2] are briefly described, and compared with earlier studies
of Jackson and Hunt [3] and Walmsley et al. [4]. A new model (FLOWSTAR) is described; it has
a more accurate description of airflow close to the surface, which is not in general logarithmic at typical measurement heights. Comparisons are made between the new model and the results of non-linear
models using higher-order turbulence schemes, especially for surface shear stress.

The widely predicted and observed drop in velocity and shear stress at the base of a dune is confirmed by FLOWSTAR. It is clear that common models for the saltation flux based only on u_* are
not appropriate at the toe of the dune where they predict a piling-up of sand.

Comparisons of the wind speed are made between the model and different sets of measurements
over a dune, by Howard and Walmsley [5], and by our group in recent field measurements over a
dune in Oman, and in a new wind-tunnel study of Howard's dune. It is found that the FLOWSTAR
calculations agree well with these sets of measurements upwind of the brink. Since the profile is not
logarithmic over the dune at the measurement heights, estimates of u_* from wind measurements
over dunes are likely to be less accurate than the FLOWSTAR computation of u_*. The saltation flux
was measured over the Oman dune and increases in proportion to computed value of u_*^3 over the dune.
This supports the use of the Lettau and Lettau version of Bagnold's flux formula for modelling sand
transport over the most of the upwind slopes of sand dunes.

1 Introduction

In the past decade, there have been many attempts to measure air flow over desert dunes
(Howard et al. [6], Knott [7], Tsoar [8], Lancaster [9], Livingstone [10], Mulligan [11]).
Meteorologists have also been studying wind flow over undulating surfaces, particularly
over hills of gentle slope in the atmospheric boundary layer. Analytical approaches have
significantly contributed to our understanding of this type of flow. The improvements in
modelling turbulent boundary layer flow have made possible the rapid calculation of the
wind field over complex terrain with gentle slope.

Howard and Walmsley [5] combined these geomorphological and meteorological approaches and found good agreement between the two, although their attempt to simulate
the development of a barchan dune was unsuccessful because of numerical instabilities.
Similar work by Wippermann and Gross [12], used the numerical simulation model FIT-
NAH (= Flow over Irregular Terrain with Natural and Anthropogenic Heat Sources)
to study the evolution of a sand dune. They started with a conical pile of sand and simulated its development and migration as a sand dune. They too found that their results were
in good agreement with observation.

These studies have shown that models of turbulent flow give good approximations to the mean flow over sand dunes and indicate how the sand flux changes. However, neither the air flow nor the sand flux calculations have been studied with the detail necessary to check whether these models are really accurate enough to predict sand dune mechanics. Certainly models of dune movement so far have not been successful in preserving the dune shape.

The prediction of sand movement over a dune requires an accurate wind-flow model in combination with a physically realistic model for the transport of sand by wind. In Howard and Walmsley's study, the turbulent flow model, MS3DJH/1.5 was based on the earlier analytical studies of air flow over low hills by Jackson and Hunt [3] (referred to here as JH) in which there is an inconsistency in matching the solution between the *inner region* and the *outer region* (as pointed out by Sykes [13]). However, the model is computationally efficient because it provides an analytical rather than iterative solution for terrain-induced flow perturbation. Wippermann and Gross [12] used the FITNAH model, in which the turbulent closure is first-order and makes use of flux-gradient relationships and mixing-length theory. Five equations were solved. Improvements on both these models have been developed by Hunt et al. [1] (referred to here as HLR), and by Zeman and Jensen [14] and Weng et al. [15].

In this study, we use the revised FLOWSTAR model of CERC (Cambridge Environmental Research Consultants Ltd), which is a computer code that runs on an IBM PC, based on the theory of HLR, to study the wind flow over barchan dunes, and its effect on the sand transport. In section 2, the FLOWSTAR model is reviewed; we describe changes which were made in order to apply the model to sand dunes. In section 3, the formulations of sand transport q and the erosion or deposition due to the divergence of q are given, and the application of the FLOWSTAR model to an idealized barchan sand dune. New measurements are reported in section 4 of wind speed (at 2 levels) and sand flux over a sand dune in Oman, and of wind speed over an idealized model barchan dune in a wind-tunnel study at Surrey University. Some preliminary results are reported here and compared with the model calculation. It is found that the agreement is satisfactory only on certain parts of a dune, lying between the 'toe' and the 'brink'; on the steeply sloping lee side of the dune the flow reverses, so that the airflow and sand flux models must be modified. One suggestion for such a change was made by Zeman and Jensen [16].

2 Turbulent airflow model

The calculations of flow characteristics over the dune are made by using FLOWSTAR. The model predicts the mean flow, streamlines and shear stress in turbulent, stratified flow over complex terrain and roughness changes. The method of calculation is to use a *Fast Fourier Transform* (FFT) algorithm for computing the Fourier transform of the terrain height and the roughness length. The algorithm for the analytical solution is then used to compute the Fourier transform of the velocity and shear stress fields. In the third stage, the transforms are inverted to calculate the actual flow variables at the required point. There is no iteration involved (as in most numerical models). Numerical methods are needed to perform the required finite Fourier transforms and Bessel function equations, but the program runs much faster than a comparable model using finite difference forms for the governing equations.

The FLOWSTAR model is based on HLR's approximate theory of flow over low hills.

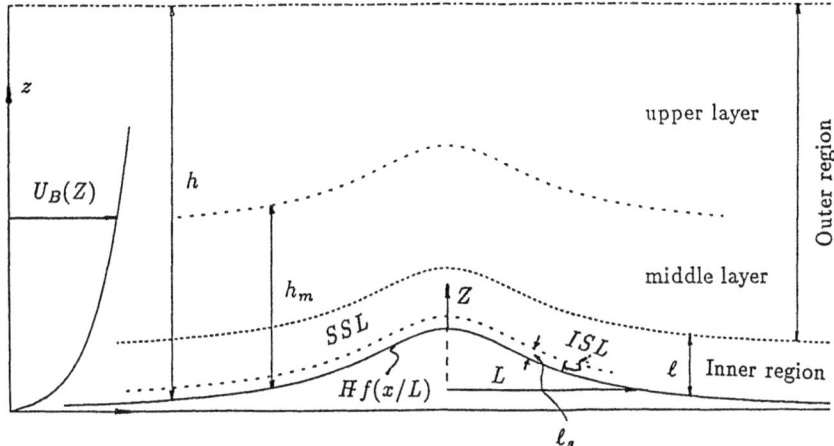

Fig. 1. Schematic diagram of flow regions over a hill for the linear analysis

The theory corrects the earlier JH solution, and extends and generalizes Sykes' solution, see [3]. Estimates of maximum speed-up and shear stress in the *outer* and *inner* regions are not much changed from the early JH theory, although there are some significant differences in the distributions of these quantities.

The model assumes that the flow can be divided into two regions according to different predominant physical processes (Fig. 1): in an *inner region*, within a distance of order l from the surface (typically l is about 1/20 of the horizontal characteristic length scale L for the terrain), the change of the shear stress significantly affects the mean flow, and the turbulence is approximately in equilibrium, so that the shear stress can be modelled by the mixing-length closure. This region can be divided into two sublayers: very close to the surface there is a very thin *inner surface layer* (ISL), in which the velocity perturbation is directly proportional to the shear stress; over most of the inner region above the ISL, within the *shear stress layer* (SSL), the shear stress perturbation decreases away from the surface and the velocity perturbation is only affected by the shear stress gradient to first order. (Analyses (see [13], [1]) show that the ISL is very thin; it can be the same order as the length of roughness elements. For present purposes the ISL may be ignored.)

Above the inner region, the changes of shear stress have negligible effects on the mean flow over the terrain. The velocity perturbation is described by inviscid theory at leading order. This is the *outer region*, which can also be divided into two sublayers: the *upper layer* which is the upper part of the outer region, in which the flow can be treated as *inviscid* and *irrotational*; the *middle layer* is the lower part of the outer region, in which the upwind shear is important and is large enough to affect the mean flow. (This layer is particularly important in stably stratified flows, where there is strong shear above the ground.)

2.1 Analysis for velocity and stress

HLR consider a steady, incompressible turbulent boundary layer flow over a single 2-D hill (ridge) with height H, a horizontal characteristic length L and the hill shape is described by $z_b(x) = Hf(x/L)$. The turbulent closure uses the *mixing-length* theory, i.e.

$$\Delta\tau = 2\varkappa u_* Z \frac{\partial \Delta u}{\partial Z}, \tag{2.1}$$

where Δu and $\Delta \tau$ are the perturbations of velocity and shear stress respectively, $\varkappa = 0.4$ is von Karman constant, and Z is the displaced-coordinate in the z-direction, which is defined as $Z = z - Hf(x/L)$. Using the matching asymptotic technique, HLR find that in the inner region the normalised perturbations of velocity and shear stress are

$$\tilde{u}_d = \frac{H/L}{U(l)} \tilde{\sigma}^{(0)} \{1 + \delta(1 - \ln \zeta - 4K_0)\}, \tag{2.2a}$$

$$\tilde{w}_d = \frac{H/L}{U(l)} \tilde{\sigma}^{(0)} \left\{ -\delta ik 2\varkappa^2 \zeta + \delta^2 2\varkappa^2 \left[2 + ik\zeta(\ln \zeta - 2) + 4\zeta \frac{\partial K_0}{\partial \zeta} \right] \right\}, \tag{2.2b}$$

$$\tilde{\tau}_d = -\frac{2H/L}{U^2(l)} \tilde{\sigma}^{(0)} \left\{ 1 + 4\zeta \frac{\partial K_0}{\partial \zeta} \right\}, \tag{2.3a}$$

and the surface shear stress perturbation is

$$\tilde{\tau}_{d0} = \frac{2H/L}{U^2(l)} \tilde{\sigma}^{(0)} \{ 1 + \delta(2\ln k + 4\gamma + 1 + i\pi) \}, \tag{2.3b}$$

where the velocity is normalised by $U_0(Z = h_m)$, the upstream velocity at the middle layer height and u_*, the upstream friction velocity. The subscript d denotes the quantities in the displaced-coordinate. For convenience, normalised coordinates are used, i.e. $\xi = x/L$ and $\zeta = Z/l$. $\delta = 1/(\ln (l/z_0)$ is a small parameter. The inner region height, l, is calculated from

$$l \ln (l/z_0) = 2\varkappa^2 L.$$

The *tilde* denotes the Fourier transform w.r.t. ξ, i.e.

$$u_d(\xi, \zeta) = \frac{1}{2\pi} \int\limits_{-\infty}^{\infty} \tilde{u}_d(k, \zeta) e^{ik\xi} dk,$$

k is the Fourier transform variable, K_0 is the modified Bessel function with argument $2\sqrt{ik\zeta}$, $\gamma = 0.57721$ is Euler's constant and $(-\sigma^{(0)})$ is the normalised pressure perturbation, calculated from

$$\sigma^{(0)}(\xi) = \frac{1}{\pi} \int\limits_{-\infty}^{\infty} \frac{f'(\xi')}{(\xi - \xi')} d\xi',$$

where $f'(\xi)$ is the derivative of $f(\xi)$ w.r.t. ξ. It is this pressure perturbation which drives the flow in the inner region.

The comparison of these solutions with numerical computations of Newley [17] (Fig. 2) and the Askervein hill experiments (Fig. 3, for experiments detail see Taylor and Teunissen [18]) show that there is a good agreement in predicting the maximum perturbation of the velocity, but the theory over-estimates the velocity perturbation in the lower part of the *inner region*, i.e. near the surface. Similarly, as $\zeta \to 0$, $\zeta \, \partial K_0/\partial \zeta \to -1/2$, the solution for shear stress perturbation (2.3a) becomes

$$\tau_d \sim \frac{2P_0}{U^2(l)} \tilde{\sigma}^{(0)}.$$

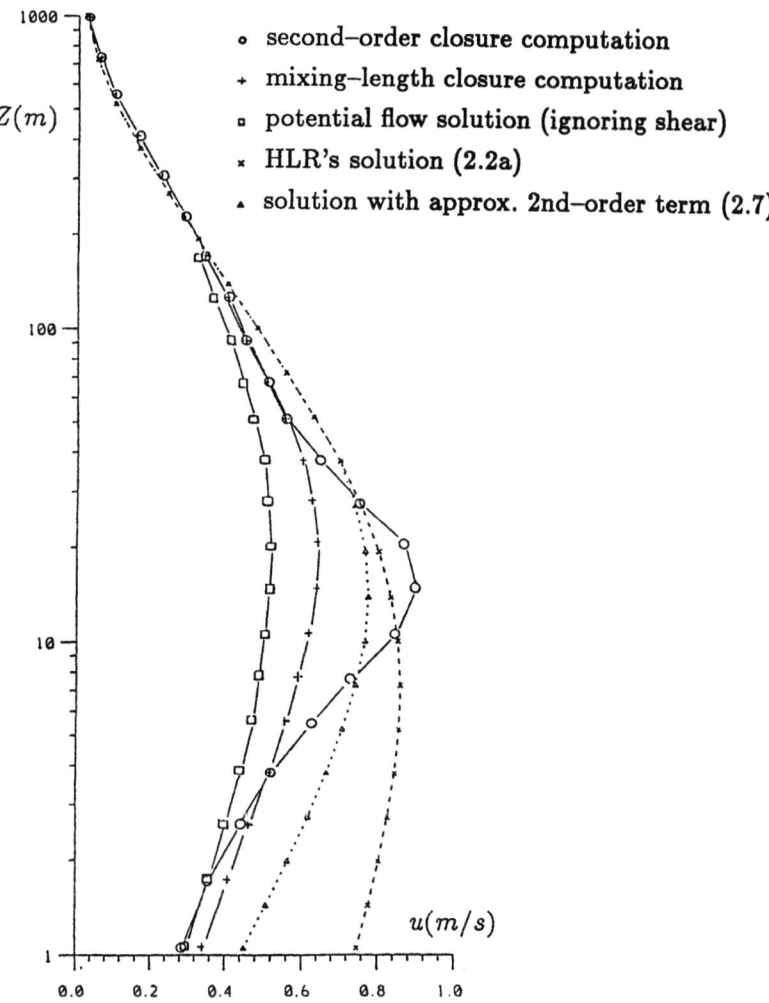

- ○ second–order closure computation
- + mixing–length closure computation
- □ potential flow solution (ignoring shear)
- × HLR's solution (2.2a)
- ▲ solution with approx. 2nd–order term (2.7)

Fig. 2. Velocity perturbation profile above the summit of a sinsoidal hill $f(x) = H + H \sin (2\pi x/\lambda)$ with $H = 20$ m, $\lambda = 2\,000$ m and $z_0 = 0.1$ m

This shows that the shear stress is only valid up to $O(\delta^0)$ in this layer. This does not match with the perturbation of shear stress at the surface given by (2.3b), which is valid up to (δ^1). For airflow over a two-dimensional Gaussian hill, $f(x) = H \exp (-x^2/L^2)$ with $H = 1\,000z_0$ and $L = 5\,000z_0$, the surface shear stress obtained by (2.3a) only gives 60% of the (2.3b) value and the maximum value occurs downstream of the crest, which is incorrect.

The reason for these large discrepancies is that the solution of HLR only satisfies the lower boundary condition asymptotically for $\ln (l/z_0) \gg 1$. As usual in the application of asymptotic theory, for a finite values of $\ln (l/z_0)$ higher order solutions are required to ensure that u_d is zero at the surface. This higher order solution is also necessary for accurate modelling of the shear stress field close to the surface, and therefore of the sand movement which is mainly confined to the inner region. In the following we use HLR's procedure to derive a new higher-order solution for both velocity and shear stress and ensure that the solution satisfies the lower boundary condition.

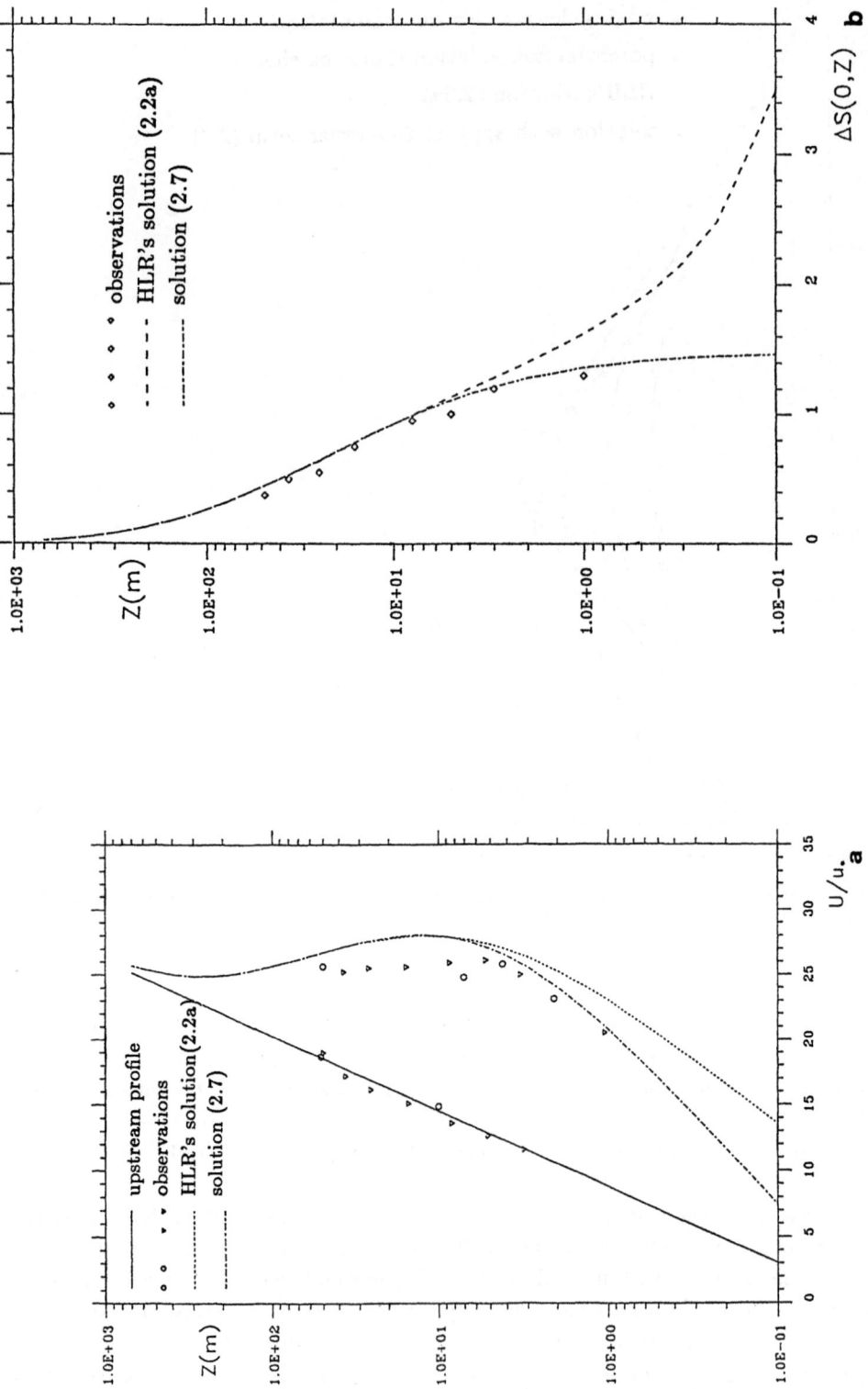

Fig. 3. Comparisons of the results from the linear model and field observations at the summit of Askervein hill (Taylor and Teunissen [18]). **a** The mean velocity. **b** The fractional speed-up

2.2 Second-order solutions

At $O(\delta^2)$, the normal shear stresses as well as the shear stresses affect the mean flow, so that they must also be considered. Since in most of the inner region, turbulence is approximately in local equilibrium, it can be assumed that the normal stresses are proportional to the Reynolds shear stress, i.e.

$$\begin{bmatrix} \tau_{xx} \\ \tau_{zz} \end{bmatrix} = \begin{bmatrix} \alpha_1 \\ \alpha_2 \end{bmatrix} \tau_{xz},$$

where the constants $\alpha_1 = 6.3$ and $\alpha_2 = 1.7$, which were derived from measurements of the atmospheric boundary layer, used by other authors, see [3]. Following the same procedure as in HLR, the second-order velocity perturbation $\tilde{u}_d{}^{(2)}$ is

$$\tilde{u}_d{}^{(2)} = \tilde{\sigma}^{(2)}(k, z_o) + \tilde{\sigma}^{(0)} \left\{ 2 - 2\varkappa^2(\alpha_1 - 2\alpha_2) - 2\ln\zeta + \ln^2\zeta + 2\varkappa^4(k^2\zeta^2 - 4 - i4k\zeta) \right.$$

$$+ 2\ln\zeta K_0 + 4\varkappa^2(\alpha_1 - \alpha_2)\left(ik\zeta K_0 - \zeta\frac{\partial K_0}{\partial\zeta}\right)$$

$$\left. + 4(2 - \ln\zeta)\zeta\frac{\partial K_0}{\partial\zeta} - 3(4\gamma + 2\ln|k| + 2/3 + i\pi)K_0 \right\}, \tag{2.4}$$

where $\tilde{\sigma}^{(2)}(k, z_0)$ is the second-order pressure perturbation at the surface. And from mixing-length theory, we obtain the following expression for $\tilde{\tau}_d{}^{(1)}$ at $O(\delta^1)$,

$$\tilde{\tau}_d{}^{(1)} = 2\tilde{\sigma}^{(0)} \left\{ 2(\ln\zeta - 1) + 4\varkappa^4(k^2\zeta^2 - i2k\zeta) + (2 + 8ik\zeta)K_0 - i4\ln\zeta k\zeta K_0 \right.$$

$$\left. + [2\ln\zeta - 4 + 4\varkappa^2(\alpha_1 - \alpha_2)ik\zeta]\zeta\frac{\partial K_0}{\partial\zeta} - 3(4\gamma + 2\ln|k| + 2/3 + i\pi)\zeta\frac{\partial K_0}{\partial\zeta} \right\}. \tag{2.5}$$

Therefore the shear stress perturbation up to $O(\delta^1)$ is

$$\tau_d = -\frac{2H/L}{U^2(l)}\tilde{\sigma}^{(0)}\left\{ 1 + 4\zeta\frac{\partial K_0}{\partial\zeta} + \delta\tau_d{}^{(1)} \right\}. \tag{2.6}$$

Jackson and Hunt [3] neglected the velocity shear and the vertical velocity in the inner region and solved the approximate equation. In their solution the perturbation velocity u_d increases with height within the inner region and its maximum is at $Z \sim l$.

With these higher-order solutions, the results are much improved (see Figs. 4 and 5). However, we can not determine the second-order pressure perturbation term $\tilde{\sigma}^{(2)}(k, z_0)$ unless the shear stress is properly modelled in the outer region. In practice, we use an approximate correction term to modify the original HLR solution, to ensure that $\tilde{u}_d \to 0$ as $\zeta \to 0$ for finite δ. By considering the complementary function $K\left(2\sqrt{ik\zeta_0}\right)$ and the feature: $K_0(x) \to 0$ as $x \to \infty$, we construct the perturbation of velocity $\tilde{u}_d$ with the approximate second-order correction as

$$\tilde{u}_d = \frac{H/L}{U(l)}\tilde{\sigma}^{(0)}\left\{ 1 - \frac{\ln\zeta - 1}{\ln(l/z_o)} - \left(2 + \frac{1}{\ln(l/z_o)}\right)\frac{K_0\left(2\sqrt{ik\zeta}\right)}{K_0\left(2\sqrt{ik\zeta_0}\right)} \right\}. \tag{2.7}$$

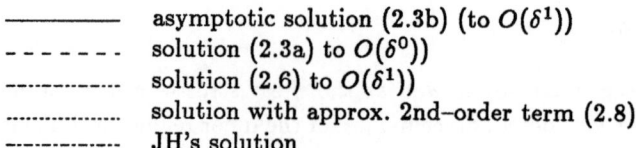

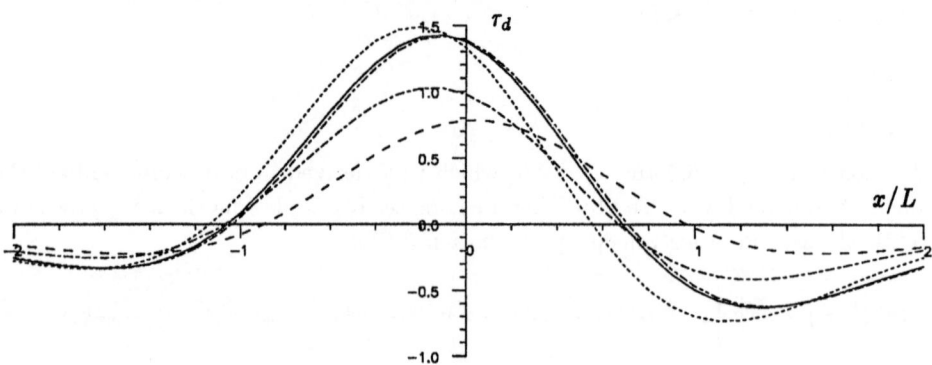

Fig. 4. Normalised surface shear stress perturbation (τ_d) for flow over a Gaussian hill, $f(x) = He^{-x^2/L^2}$, with $H = 1\,000z_0$ and $L = 5\,000z_0$

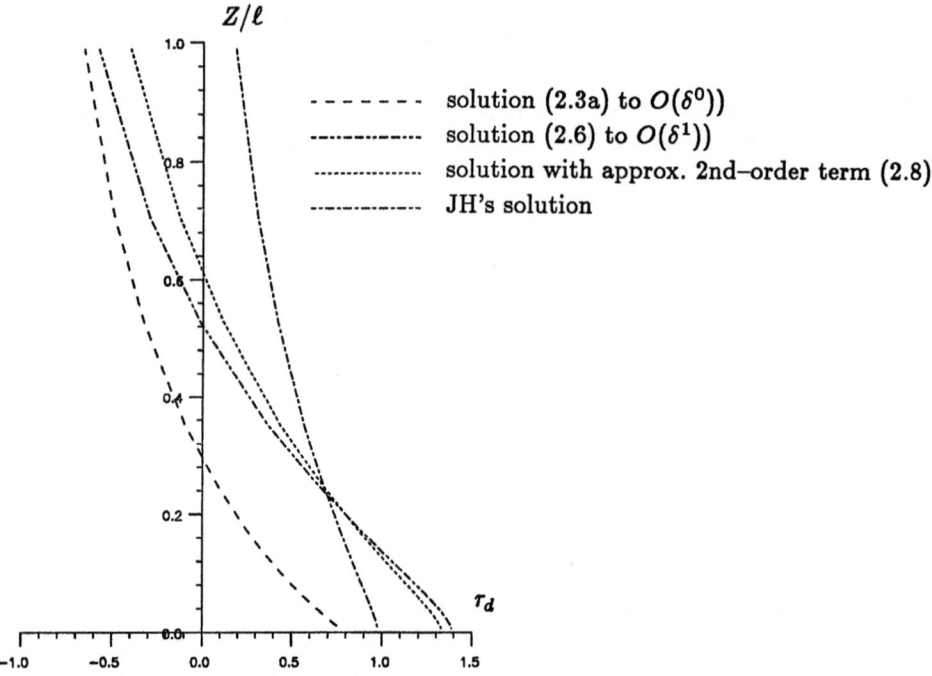

Fig. 5. Normalised shear stress perturbation (τ_d) above the summit of a Gaussian hill, $f(x) = He^{-x^2/L^2}$, with $H = 1\,000z_0$ and $L = 5\,000z_0$

(The terms in $K_0\left(2\sqrt{ik\zeta}\right)$ are similar to the solution of JH). At $\zeta = \zeta_0(= z_0/l)$, the solution gives exactly $\tilde{u}_d = 0$. From this solution, we can also obtain the perturbation of shear stress with the approximate first-order correction term, i.e.

$$\tilde{\tau}_d = -\frac{2H/L}{U^2(l)}\,\tilde{\sigma}^{(0)}\left\{1 + [2\ln(l/z_0) + 1]\frac{\zeta\,\partial K_0/\partial\zeta}{K_0\left(2\sqrt{ik\zeta_0}\right)}\right\}. \tag{2.8}$$

As $\zeta \to 0$, i.e. very small ζ, it reduces to

$$\tilde{\tau}_d \approx \frac{2H/L}{U^2(l)}\,\tilde{\sigma}^{(0)}\left\{1 + \frac{2\ln|k| + 4\gamma + 1 + i\pi}{\ln(l/z_0)}\right\}. \tag{2.9}$$

This is equal to the asymptotic solution for the surface shear stress which HLR obtained.

Comparison of these approximate expressions with both the solutions including the *exact* second-order term and experimental data shows good agreement, Fig. 2 and 3. By including the approximate second-order term, the result is much improved, particularly near the surface. The comparisons of perturbation of shear stress are shown in Fig. 4 and 5. Again the solution with the approximate second-order term agrees well with both the asymptotic solution (2.9b) and the solution with *exact* second-order solution (2.6).

To test our formula for the surface shear stress, we compared the results of linear theory with the numerical computation (Weng [19]). In the numerical calculation, the Reynolds shear stress is assumed to be proportional to the local mean velocity gradient and an eddy diffusivity K, where K is assumed to be proportional to the square root of the turbulent kinetic energy E and a turbulent length scale l_m. Two length scale formulae were used. One was the usual mixing-length scale (OML) (which is widely used eg. by Taylor and Gent [20], Taylor [21], [22] and Mason and King [23])

$$l_m^{-1} = \frac{1}{\lambda_B} + \frac{1}{\varkappa(z + z_0)}, \tag{2.10}$$

and the other was the *shear-blocking mixing length* (SBML):

$$l_m^{-1} = \frac{1}{\lambda_B} + \frac{A_s\langle\partial U/\partial z\rangle}{\langle w'\rangle} + \frac{A_B}{(z + z_0)}, \tag{2.11}$$

where λ_B is the maximum value obtained l_m for large z, the angle brackets the average value over a typical addy scale and the shear and blocking coefficients A_s and A_B can be determined by comparing (2.11) with measurements in different kinds of neutrally stable flows near boundaries. They are $A_s \approx 1.0$ and $A_B \approx 0.6$. The vertical fluctuation velocity w' is calculatedf rom $w' = 1.3\tau$, where τ is the local shear stress.

Formula (2.10) prescribes that the turbulent length scale is proportional to the distance from the lower surface. Experiments show that there are very high shear regions just downwind of the summit of hills. Britter et al. [24] suggested that the turbulent length scale must be smaller than that predicted by (2.10) in this region; and is controlled by *high shear*. In the new formula (2.11), the effects of both *shear* and *blocking* of the boundary are included. The numerical simulation results of boundary layers and channel flow have proved the

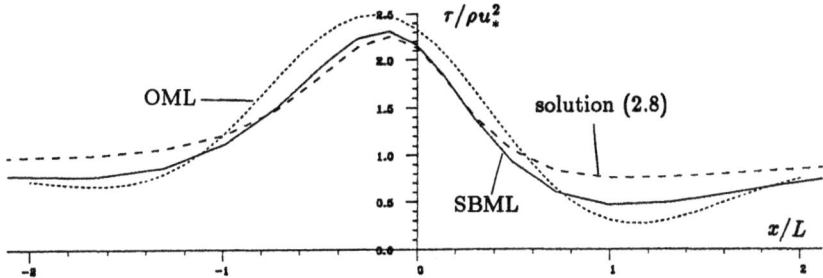

Fig. 6. Comparison of normalised surface shear stress between the linear theory and the numerical computations for flow over a Gaussian hill, $f(x) = He^{-x^2/L^2}$, with $H = 1000z_0$ and $L = 5000z_0$

validity of this *new* lengthscale (Spalart [25], [26], Mansour et al. [27]). The comparison of surface shear stress by the linear theory and the numerical computation of the mixing-length model are shown in Fig. 6. The solution with the approximate second-order term (2.8) agrees well with the numerical computation of the SBML model.

2.3 Solution for three-dimensions

The solutions for the airflow which we have described are easy to extend to three dimensions. We use them as the basis of our calculation of the wind flow over sand dunes.

In the three-dimensional case, the normalised perturbations of velocity for the x- and y-component in the *inner region* are:

$$\tilde{u}_d = \frac{\tilde{f}(k_1, k_2)\, k_1{}^2}{k_{12}}\, e^{-k_{12}Z}\, \frac{1}{U(l)} \left\{ 1 - \frac{\ln \zeta - k_{12}^2/k_1{}^2}{\ln (l/z_0)} - \left[2 + \frac{k_{12}^2/k_1{}^2}{\ln (l/z_0)} \right] \frac{K_0}{K_0\big(2\sqrt{ik\zeta_0}\big)} \right\}, \quad (2.12\,\text{a})$$

$$\tilde{v}_d = \frac{\tilde{f}(k_1, k_2)\, k_1 k_2}{k_{12}}\, e^{-k_{12}Z}\, \frac{1}{U(l)} \left\{ 1 - \frac{4K_0\big(2\sqrt{2ik_1\zeta}\big)}{\ln (l/z_0)} \right\}, \quad (2.12\,\text{b})$$

and the normalised perturbations for shear stress in the x- and y-directions are:

$$\tilde{\tau}_{dx} = -\frac{\tilde{f}(k_1, k_2)\, k_1{}^2}{k_{12}}\, e^{-k_{12}Z}\, \frac{2}{U^2(l)} \left\{ 1 + \left[2\ln (l/z_0) + k_{12}^2 k_1{}^2\right] \frac{\zeta\, \partial K_0/\partial \zeta}{K_0\big(2\sqrt{ik\zeta_0}\big)} \right\}, \quad (2.13\,\text{a})$$

$$\tilde{\tau}_{dy} = -\frac{\tilde{f}(k_1, k_2)\, k_1 k_2}{k_{12}}\, e^{-k_{12}Z}\, \frac{4\zeta\, \partial K_0\big(2\sqrt{2ik_1\zeta}\big)/\partial \zeta}{U^2(l)}, \quad (2.13\,\text{b})$$

where $\tilde{f}(k_1, k_2)$ is the two-dimensional Fourier transform of terrain height and k_1 and k_2 are Fourier transform variables, representing the x- and y-directions respectively and $k_{12} = \sqrt{k_1{}^2 + k_2{}^2}$.

Similarly, the perturbation of the surface shear stress in the x- and y-direction can be written as

$$\tilde{\tau}_{dx0} = \frac{\tilde{f}(k_1, k_2)\, k_1{}^2}{k_{12}}\, \frac{2}{U^2(l)} \left\{ 1 + \frac{2\ln |k_1| + 4\gamma + 1 + i\pi}{\ln (l/z_0)} \right\}, \quad (2.14\,\text{a})$$

$$\tilde{\tau}_{dy0} = \frac{\tilde{f}(k_1, k_2)\, k_1 k_2}{k_{12}}\, \frac{2}{U^2(l)}. \quad (2.14\,\text{b})$$

where the subscript 0 denotes the surface value.

These *inner region* solutions have been incorporated, with the other solutions of HLR, into the FLOWSTAR model. For model details see [2].

3 Airflow, sand transport and erosion rate over an idealized barchan dune

3.1 Airflow over an idealized barchan dune

The barchan dune we first study here is based on the one studied by Howard and Walmsley [5] and Howard et al. [6] (it was originally near the Salton Sea, California, referred to here as Howard's dune). In order to run the FLOWSTAR model, the basic inputs are terrain

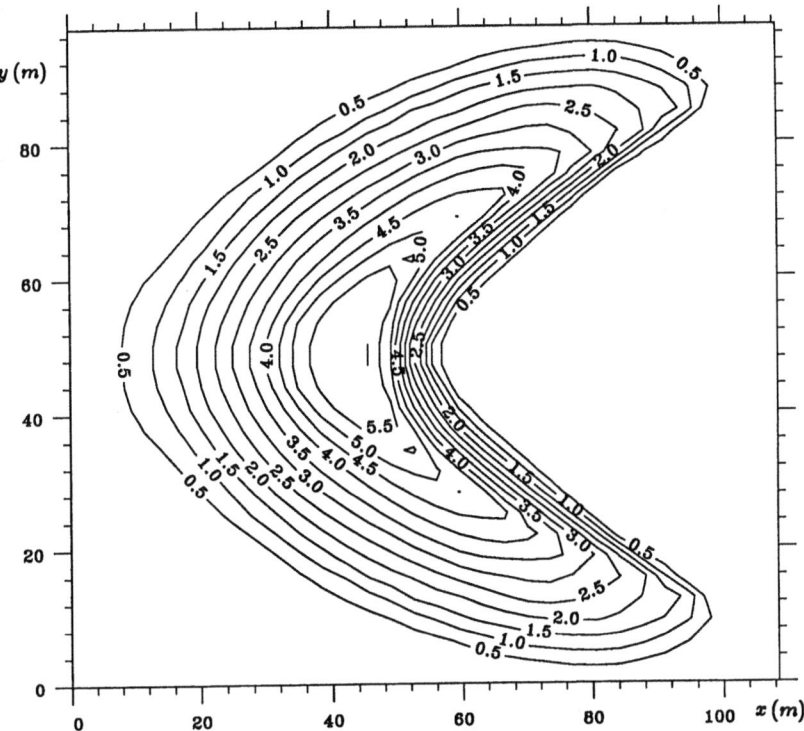

Fig. 7. Contour map of revised Howard's barchan dune

(dune) data, roughness data, the upstream wind direction, the upstream velocity profile and the heights at which mean wind is calculated. The input terrain data was digitised from a symmetrised and smoothed version of the Howard's dune. Figure 7 shows the contour map of this dune. It is about 108 m long (x-direction), 97 m wide (y-direction) and 6 m high (z-direction).

Selection of an appropriate value of z_0 is restricted by a lack of relevant field data. z_0 is partially determined by fixed field elements, such as rocks, pebbles and shrubs on the desert bed and ripples on the dune, and partly by the saltating sand; this is important during periods of high wind velocity when most sand transport occurs, and makes measurement difficult.

Because of the non-uniform distribution of roughness length z_0 over the dune, the wind speed profile is modified from the usual logarithmic profile. Bagnold [28] suggested

$$U(z) = \frac{u_*}{\varkappa} \ln \frac{z}{z_r} + U_t,$$

where u_* is spatially varying friction velocity, z_r is a constant reference height and U_t is the threshold speed for sand movement at that height, known as the *impact threshold velocity*. It is calculated from

$$U_t = \frac{A}{\varkappa} \left(\frac{g d \varrho_s}{\varrho} \right)^{1/2} \ln \frac{z_r}{z_0},$$

where z_0 is the roughness length of a completely flat sand surface ($z_0 = 0.002$ cm for $d = 0.25$ mm), z_r/z_0 is of the order of magnitude 100, A is a constant with a value of about

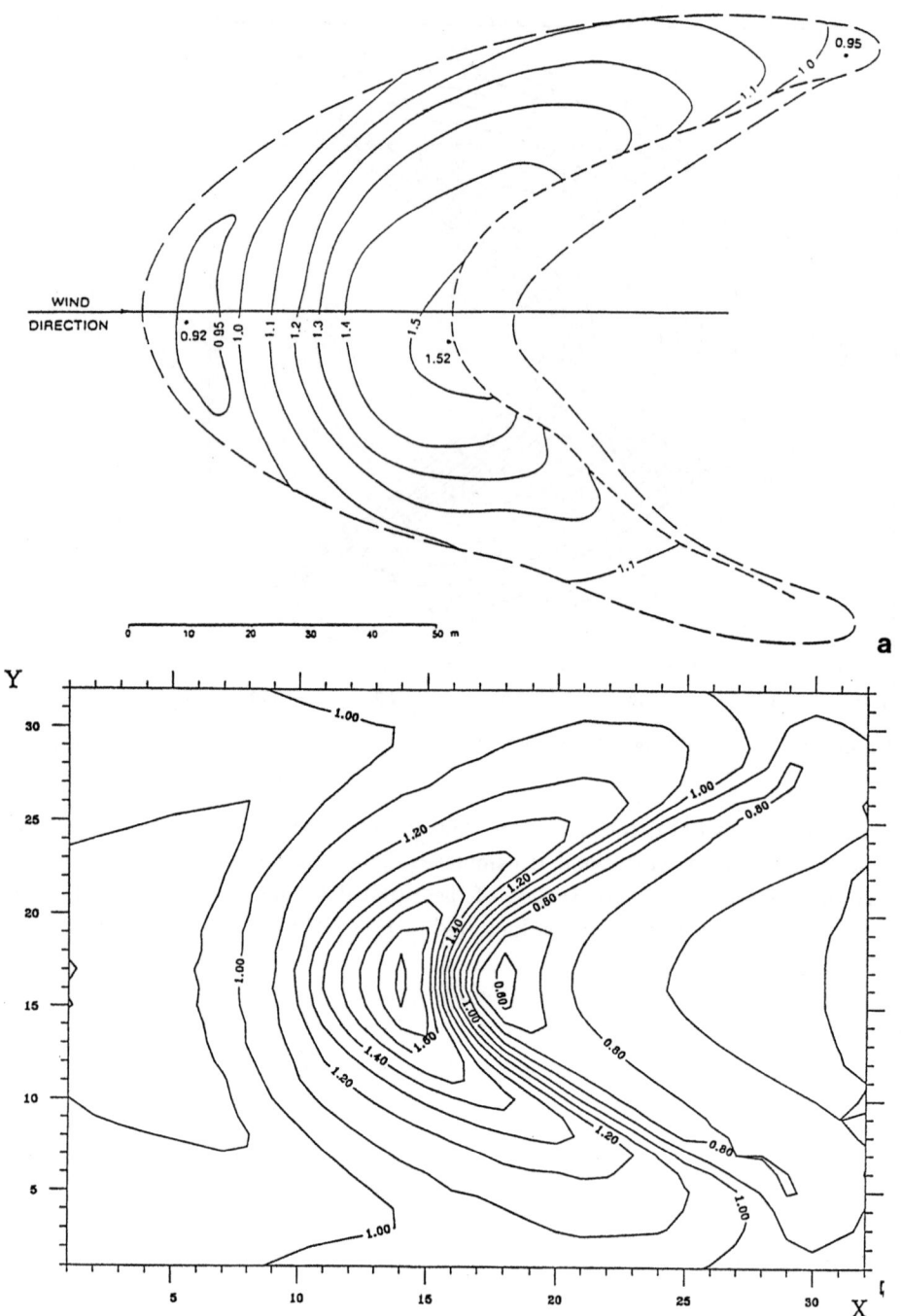

Fig. 8. Normalised wind speed at 80 cm above the terrain surface. Contours show values $\sqrt{U^2 + V^2}/U_0$. Contour interval is 0.1. **a** Field observation (Howard et al. [6]), **b** FLOWSTAR model calculation (X and Y in grid units)

0.1 for air, ϱ_s is the bulk density of sand, ϱ the density of air, g the gravitational acceleration and d the grain size (see also Greeley and Iversen [29]). It is assumed that the saltation flux is low enough on the desert floor for z_0 to be determined only by the immobile surface.

In the following calculations, the upstream velocity profile is assumed to be logarithmic with the friction velocity $u_* = 0.35\,\mathrm{ms^{-1}}$ and the uniform roughness distribution $z_0 = 0.001\,\mathrm{m}$. The stratification is neutral, the upstream wind direction is taken to be in the x-direction, and the calculation grid size is 32×32.

The comparison of the FLOWSTAR model computation with field and wind-tunnel observations of Howard et al. [6] for normalised wind speed at 80 cm above the terrain surface is shown in Fig. 8. Since the empirical observations were only made at a few (13) points over the dune, it is preferable to compare each data point with the computed values. Note that our model calculation gives us a maximum speed-up (defined as the mean wind speed $\sqrt{U^2 + V^2}$ divided by the upstream value U_0) of about 1.8 at this height. The maximum measured value is 1.52 at a point where the model predicted 1.6. This is a satisfactory level of agreement. It is expected that the model slightly overpredicts wind speed because it does not allow for the separated air flow in the wake and it is linear.

3.2 Sand transport

Assuming that the sand transport q is largely by saltation, we use Lettau and Lettau's [30] formula for the sand flux over level terrain:

$$q = \begin{cases} C(\varrho/g)\, u_*^{\,2}(u_* - u_{*t}), & u_* > u_{*t}; \\ 0, & u_* \leqq u_{*t}, \end{cases} \tag{3.1}$$

where u_* is the local shear velocity, u_{*t} is the so called *threshold shear velocity* and C is a constant. Lettau and Lettau found

$$C = 5.5, \quad \text{and} \quad u_{*t} = 0.22\,\mathrm{ms^{-1}}.$$

This is for the two dimensional case. It is easy to extend to three-dimensions, eg.

$$\boldsymbol{q} = \begin{cases} C(\varrho/g)\, \boldsymbol{u}_* \,|\boldsymbol{u}_*|\,(|\boldsymbol{u}_*| - |\boldsymbol{u}_{*t}|), & |\boldsymbol{u}_*| \geqq |\boldsymbol{u}_{*t}|; \\ 0, & |\boldsymbol{u}_*| < |\boldsymbol{u}_{*t}|. \end{cases} \tag{3.2}$$

It is this formula which we have used to compute sand transport.

Howard et al.'s comparisons of model with the experimental data showed that agreement was better by not including slope effects in the transport formula, and was not improved by including spatial lag of transport following changing u_*. Therefore, in this study, we also ignore slope effects and spatial lags.

The height $h(x, y, t)$ of the surface of a dune varies with time if there is net erosion or deposition of sand. This is associated with a reduction or increase in sand transport per unit length in the direction of the sand transport. This is defined mathematically as $\mathrm{div}\,\boldsymbol{q} = \partial q_x/\partial x + \partial q_y/\partial y$ (where q_x and q_y are the sand transport in the x- and y-directions respectively). As long as the slope of the surface is small and less than the slope of the slip-face, then the rate of increase of h is given in terms of $\boldsymbol{q}$ by

$$\varrho_s \frac{\partial h}{\partial t} = -\mathrm{div}\,\boldsymbol{q}, \quad \text{if} \quad |\nabla h| \ll 1. \tag{3.3}$$

The sand transport equation (3.1) and its divergence $\mathrm{div}\,\boldsymbol{q}$ can be easily computed using a *Fast Fourier Transform* method (Press et al. [31]). Using a program similar to FLOWSTAR, these calculations are performed in the program called SANDTR, which runs on an IBM PC. Fig. 9 shows the contours of the divergence of q, i.e. div q for the Howard's dune, in which the positive and negative values represent erosion and deposition respectively (in this calculation the threshold shear velocity is assumed to be $u_* = 0.22\,\mathrm{ms^{-1}}$).

There are some interesting implications of equation (3.3), which have not all been thoroughly investigated. For example as the flow approaches an isolated dune, it slows down (Tsoar [32], Livingstone [10]), i.e. $\partial u/\partial x < 0$, and it diverges laterally, i.e. $\partial v/\partial y > 0$. Also it moves upwards, i.e. $\partial w/\partial z > 0$, but $\partial u/\partial x + \partial v/\partial y = -\partial w/\partial z < 0$. Therefore, since $q_x \propto U + u, q_y \propto v$,

$$\frac{\partial q_x}{\partial x} + \frac{\partial q_y}{\partial y} < 0, \tag{3.4}$$

and so from (3.3), $\partial h/\partial t$ is predicted to be positive just upwind of the dune. This is not generally observed, perhaps because sand transport *on* the dune is different and greater than that on the less mobile level upwind, i.e. the upwind flow may not be saturated with sand, or because there is three-dimensional divergence of transport, or because high turbulence is causing lift. If the last of these hypotheses is correct the sand transport equation would need revision.

From (3.3) it is also possible to examine how a sand dune moves and whether it changes its shape as it moves. If a dune moves at a speed U_d *without change of shape*, then

$$\frac{\partial h}{\partial t} = -U_d \frac{\partial h}{\partial x}. \tag{3.5}$$

In Fig. 10 contours of $\partial h/\partial x$ for the revised Howard's dune is plotted. Note that they are similar to the contours of $-\mathrm{div}\, q$ or $\partial h/\partial t$, but there are some differences in details. Never-

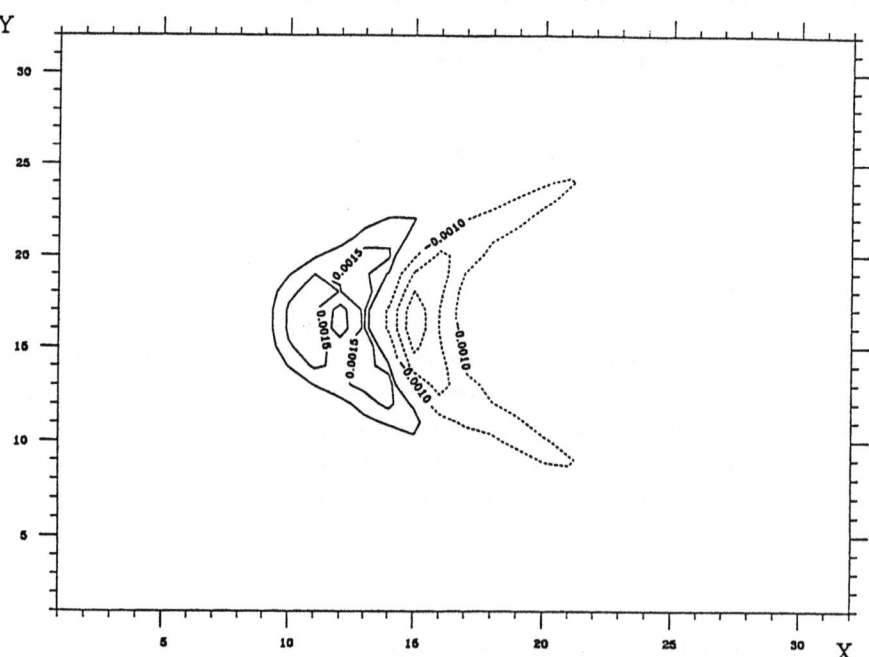

Fig. 9. Convergence/divergence of sand transport (div q/ϱ). Contour interval 0.0005 (X and Y in grid units)

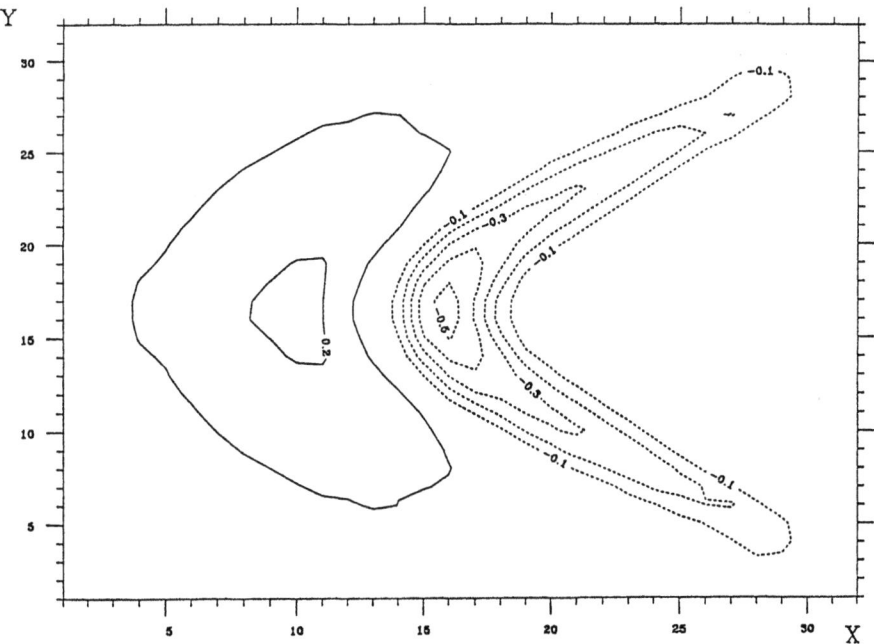

Fig. 10. Derivative of terrain height w.r.t. x $(\partial H/\partial x)$. Contour interval 0.1 (X and Y in grid units)

theless it is plausible that (3.3) and (3.5) might be a first order approximation to a model of dune movement, as has been assumed by several previous authors.

For the case of Howard's dune, taking $|\mathrm{div}\, \boldsymbol{q}|_{\max} \approx 1.5 \times 10^{\,0}$ from Fig. 9, if the dune does not change shape asit moves downwind, we can estimate that the change of the dune surface height is, $\partial h/\partial t \sim -5$ mm/hr (here we have used the air-to-packed sand density ratio, $\varrho/\varrho_s \approx 10^{-3}$). From the Fig. 10, $\partial h/\partial x \approx 0.1$, we find that the dune moves forward at speed $U_d \sim 5$ cm/hr, which is similar to the estimate by Zeman and Jensen [16].

4 Comparison of experiments in the field and wind-tunnel with FLOWSTAR calculations

The wind-tunnel studies were carried out at the Department of Mechanical Engineering, Surrey University. The tunnel was a blower type with a working section of $4.0 \times 0.9 \times 0.75$ metres. The model dune is scaled at $1:200$ assuming the full scale dune is similar to a symmetrised version of Howard's barchan dune and has a height of 6 metres (Fig. 7). The undisturbed boundary layer was simulated by using a mixing mesh, fence and roughness as developed by Cook [33]. The desert roughness was reproduced by using Lego baseboard.

The velocity and turbulence measurements were made using a DISA constant temperature hot-wire anemometer system, employing both single and cross-wires. The upstream velocity profile was found to be logarithmic, $U(z) = (u_*/\varkappa)\ln\big((z-d)/z_0\big)$ with the friction velocity $u_* = 0.4546\ \mathrm{ms}^{-1}$, von Karman constant $\varkappa = 0.4$, roughness length $z_0 = 0.06$ mm and the displacement $d = -1.5$ mm.

Figure 11 shows the comparisons of vertical velocity profile at four different positions over the model dune. There is good agreement. But near the crest the FLOWSTAR calculation overpredicts wind speed, because no allowance in model for the wake region. Figure 12

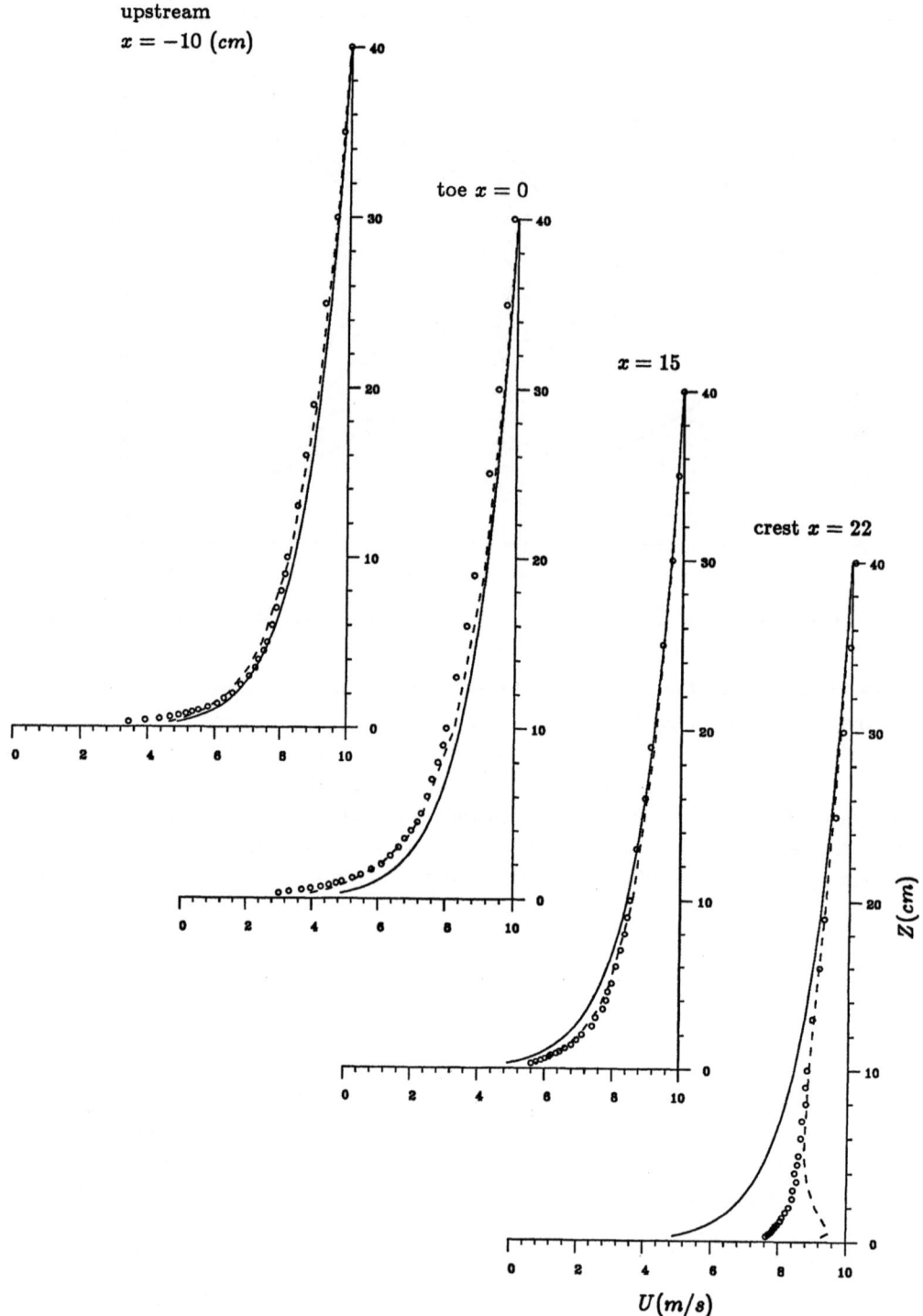

Fig. 11. Comparison of wind speed profile at four different positions over the model dune. Symbols are wind-tunnel measurements, solid line upstream profile and dashed lines the FLOWSTAR calculation

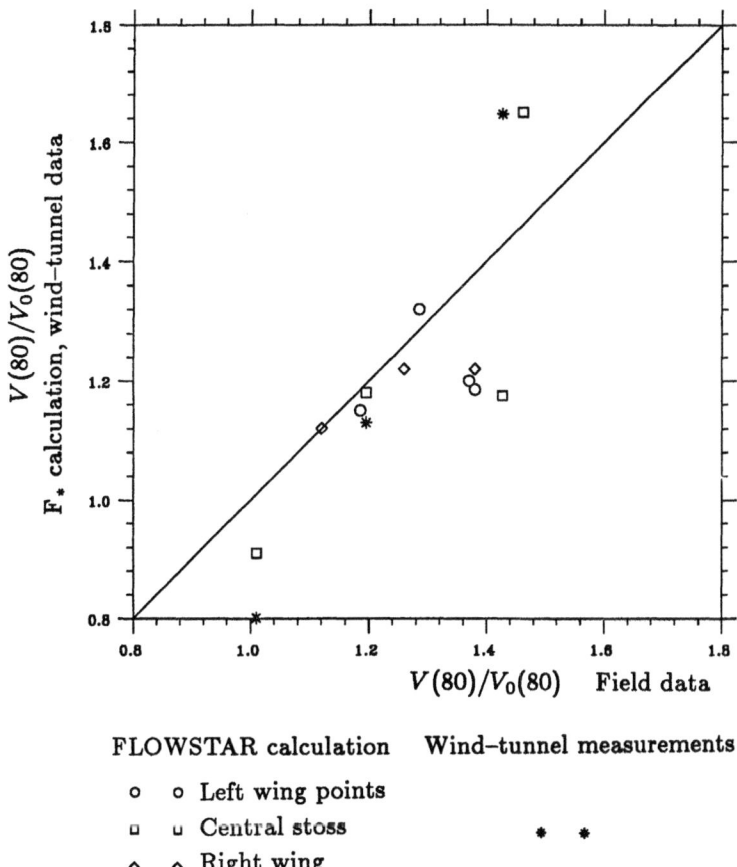

FLOWSTAR calculation   Wind–tunnel measurements

  o   o Left wing points

  □   □ Central stoss   *   *

  ◇   ◇ Right wing

Fig. 12. Comparisons of normalised wind speed measured in the field (Howard et al. [6]), FLOWSTAR calculation and wind-tunnel measurements over model dune at equivalent heights

shows comparisons between the normalised velocities measured in the field (see [6]), the FLOWSTAR calculation and wind-tunnel measurements over the model dune (at Surrey University) at the same relative locations. The differences are likely to be due to the large natural variance in the measurements both in the field and wind-tunnel.

Field measurements were carried out on three barchan sand dunes in eastern Oman between Ashkharah and Ras-al-Hadd in August, 1989.

The 1 minute averaged wind speed was measured and surface shear stress calculated with an array of 26 cup-anemometers set up along the centre-line of the dune at heights of 0.25 m and 0.35 m. The data from these anemometers were recorded by Grant Data Loggers downloaded onto a Toshiba 1200 portable computer.

Sand traps were placed next to each anemometer station in order to measure the sand flux. The sand traps were made from sections of double polycarbonate sheeting, 20 cm high. The amount of sand collected by each of the eight sections during an exposure period of 2 minutes (for crest areas) or 9 minutes (for toe areas) was measured.

The upstream reference velocity was measured by an array of anemometers positioned 30 metres upwind of the dune centre-line. These recorded both wind velocity and direction every 5 minutes at four heights up to a reference anemometer at 3.4 metres. The upstream roughness was calculated by fitting the logarithmic velocity profile to the data obtained at this meteorological station. This gives $z_o = 3$ mm with sand in saltation.

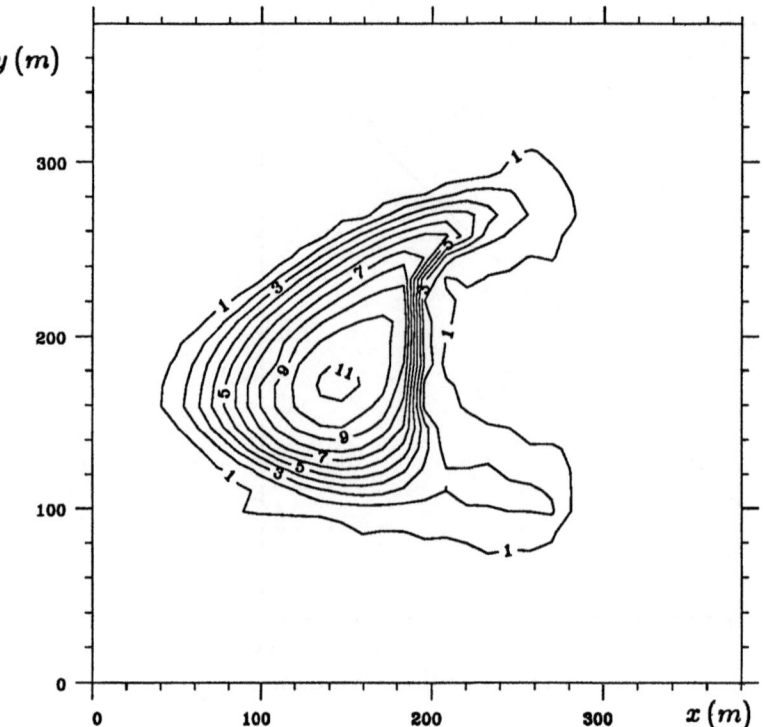

Fig. 13. Contour map of Caliph barchan sand dune

Figure 13 shows the contour map of one barchan dune with height about 11 metres and length 150 metres from toe to brink line. The comparison of FLOWSTAR calculation with the field measurements on this dune is shown in Fig. 14a (the toe of the dune is at $x = 0$ and the crest of dune is at $x = 105$ m). The experimental data were all taken on the 22nd of August, 1989, but at different times. To make the velocity measurements comparable, they have been normalised by the velocity at the meteorological station. For each point, the presented velocities are averaged over 20 minutes. The input upstream profile for the FLOWSTAR calculation is logarithmic with the roughness length $z_0 = 3$ mm (as measured), the friction velocity $u_* = 0.5114$ ms^{-1} and the wind speed at the reference height $z = 3.42$ m is 9 ms^{-1}. The friction velocity is calculated by assuming that the upstream velocity is logarithmic and taking the average velocity as that measured at the meteorological station for the period of measurements of velocities over the dune (at 12 points). The agreement is encouraging, but the model calculation seems to underpredict the wind strength.

The comparison of the shear velocity profile along the centre-line up to the brink line of the field dune is shown in Fig. 14b. u_* values were calculated from the vertical velocity profile estimated from the measurements at two levels, assuming it to be logarithmic. The data are very scattered in the range from 0.3786 ms^{-1} to 1.2825 ms^{-1}. It is likely that between these measured heights ($z = 0.25$ m and 0.35 m), the velocity profile is *not* logarithmic. The FLOWSTAR calculation gives u_* in the range from 0.3937 ms^{-1} to 0.9643 ms^{-1}. We can also obtain the shear velocity from the model's calculations of velocities at two levels again assuming that the velocity profile is logarithmic, these values show small variations.

The computed value of $u_*{}^3$ is found to be proportional to the measured saltation sand

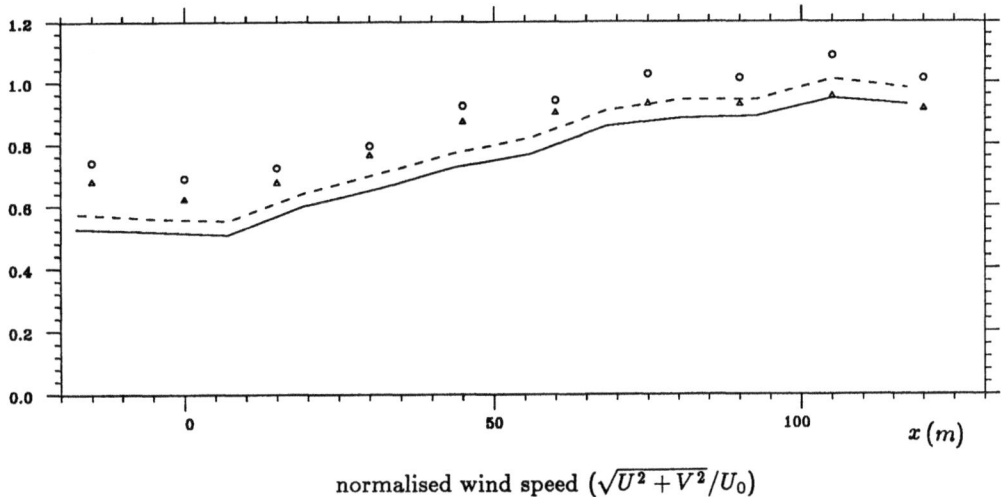

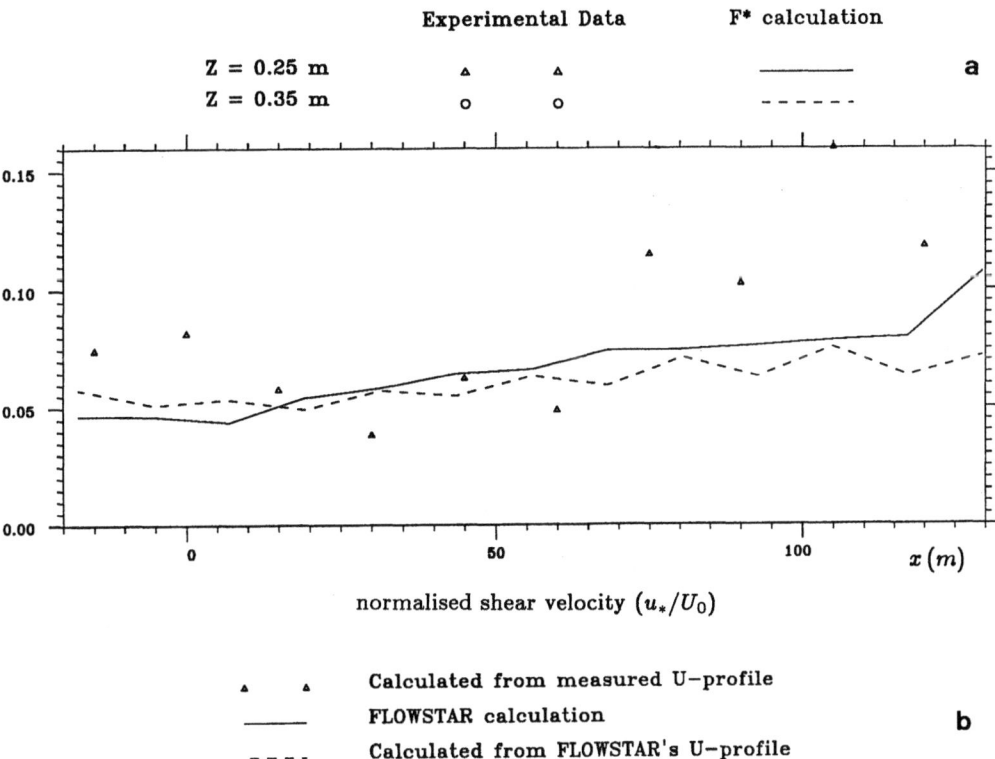

Fig. 14. Comparisons of FLOWSTAR calculation with the field measurements along the centre-line of the Caliph dune. **a** normalised wind speed, **b** normalised shear velocity

flux over the dune in Oman, as shown by the sand trap data. Figure 15 shows the normalised sand volume (S_v/U_r) and the calculated value of $(u_*/U_0)^3 \times 10^4$ (where U_0 is the reference velocity at 3.42 m) at 'toe' $(x = 0)$ and 'crest' $(x = 105$ m) areas. There is good agreement between these two. It supports the use of Bagnold's flux formula for modelling sand transport over upwind slopes of sand dunes (section 3).

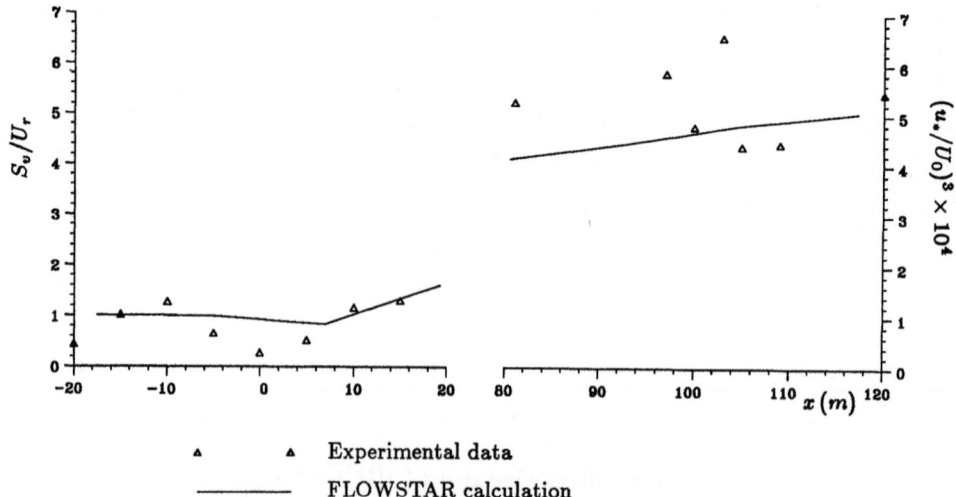

Fig. 15. Normalised the measured sand volume (S_v) with the upstream reference velocity U_r and the computed value of $(u_*/U_0)^3 \times 10^4$ by the FLOWSTAR model

Note that in the field experiment the roughness lengths both upwind and over the dune are similar, but in the wind-tunnel experiment there is change in roughness length — over the model dune it is smooth and upstream is rougher. More recent experiments with a rough model dune have yet to be fully analysed.

5 Conclusion

In this paper we have presented new developments in mathematical modelling of air flow over sand dunes and a comparison with wind-tunnel and field observations of air flow and sand flux over barchan dunes. Preliminary results are encouraging, although further questions have been raised. One concerns the behaviour of the sand flux at the base of the windward slope, where the flow is known to decelerate, the other concerns discrepancies between field and model predictions of wind speed over the dunes which may be reduced with more detailed field measurement. The third concerns the poor modelling of flow on the lee side of the barchan dunes.

Acknowledgements

This project was financed by NERC and CERC Ltd. We are grateful to Sultan Qaboos University and the Diwan of Royal Court Affairs of the Sultanate of Oman, for their help during the field experiments and to CERC for the use of its computing facilities.

References

[1] Hunt, J. C. R., Leibovich, S., Richards, K. J.: Turbulent shear flow over hills. Quart. J. R. Met. Soc. 114, 1435−1470 (1988).
[2] Carruthers, D. J., Hunt, J. C. R., Weng, W. S.: A Computational model of stratified turbulent air flow over hills − FLOWSTAR I. Proc. of Envirosoft. In: Computer techniques in environmental studies (ed. Zanetti) Berlin, Heidelberg, New York: Springer, pp. 481−492 (1988).

[3] Jackson, P. S., Hunt, J. C. R.: Turbulent wind flow over a low hill. Quart. J. R. Met. Soc. **101**, 929—955 (1975).

[4] Walmsley, J. L., Salmon, J. R., Taylor, P. A.: On the application of a model of boundary layer flow over low hills and real terrain. B. L. Met. **23**, 17—46 (1982).

[5] Howard, A. D., Walmsley, J. L.: Simulation model of isolated dune sculpture by wind. In: Proc. Inter. Workshop on the Physics of Blown Sand, Memoirs No. 8. Department of Theoretical Statistics, University of Aarhus, Aarhus, 377—391 (1985).

[6] Howard, A. D., Morton, J. B., Gad-el-Hak, M., Pierce, D. B.: Sand transport model of barchan dune equilibrium. Sedimentology **25**, 307—338 (1978).

[7] Knott, P.: The structure and pattern of dune-forming winds. Ph. D thesis, University of London (1979).

[8] Tsoar, H.: Dynamic processes acting on a longitudinal (seif) sand dune. Sedimentology **30**, 567 to 578 (1983).

[9] Lancaster, N.: Variations in wind velocity and sand transport on the windward flanks of desert sand dunes. Sedimentology **32**, 581—593 (1985).

[10] Livingstone, I.: Geomorphological significance of wind flow patterns over a Namib linear dune. In: Aeolian geomorphology. Binghampton Symposia in Geomorphology, International Series, **17**, (ed. Nickling, W. G.) Boston: Allen and Unwin, pp. 97—112 (1986).

[11] Mulligan, K. R.: Velocity profiles on the windward slope of a transverse dune. Earth Surface Process and Landforms **13**, 573—582 (1987).

[12] Wippermann, F. K., Gross, G.: The wind-induced shaping and migration of an isolated dune: a numerical experiment. B. L. Met. **36**, 319—334 (1986).

[13] Sykes, R. I.: An asymptotic theory of incompressible turbulent boundary-layer flow over a small hump. J. Fluid Mech. **101**, 647—670 (1980).

[14] Zeman, O., Jensen, N. O.: Modifications to turbulence characteristics in flow over hills. Quart. J. R. Met. Soc. **113**, 55—80 (1987).

[15] Weng, W. S., Richards, K. J., Carruthers, D. J.: Some numerical studies of turbulent wake over hills. In: Advances in turbulence 2. (eds. Fernholz, H.-H., Fiedler, H. E., Berlin, Heidelberg: Springer, pp. 123—456 (1989).

[16] Zeman, O., Jensen, N. O.: Progress report on modelling permanent form sand dunes. Risø National Laboratory, Roskilde, Denmark (1988).

[17] Newley, T. M. J.: Turbulent air flow over hills. Ph. D thesis, University of Cambridge (1986).

[18] Taylor, P. A., Teunissen, H. M.: The Askervein hill project: report on the September October 1983 main field experiment. Internal report MSRB-84-6, Atmos. Envir. Service, Downsview, Ontario, Canada (1985).

[19] Weng, W. S.: Turbulent air flow and fluxes over low hills. Ph. D thesis, University of Cambridge (1989).

[20] Taylor, P. A., Gent, P. R.: A model of atmospheric boundary-layer flow above above an isolated two dimensional "hill"; an example of flow above "gentle topography". B. L. Met **7**, 349—362 (1974).

[21] Taylor, P. A.: Some numerical studies of surface boundary-layer flow above gentle topography. B. L. Met. **11**, 439—465 (1977).

[22] Taylor, P. A.: Numerical studies of neutrally stratified planetary boundary layer flow above gentle topography. B. L. Met. **12**, 37—60 (1977).

[23] Mason, P. J., King, J. C.: Atmospheric flow over a succession of nearly two dimensional ridges and valleys. Quart. J. R. Met. Soc. **110**, 821—845 (1984).

[24] Britter, R. E., Hunt, J. C. R., Richards, K. J.: Air flow over a 2-dimensional hill: studies of velocity speed-up, roughness effects and turbulence. Quart. J. R. Met. Soc. **107**, 91—110 (1981).

[25] Spalart, P. R.: Numerical study of sink flow boundary layer. J. Fluid Mech. **172**, 307—328 (1986).

[26] Spalart, P. R.: Direct simulation of a turbulent boundary layer up to $R_\theta = 1410$. J. Fluid Mech. **187**, 61—98 (1986).

[27] Mansour, N. N., Kim, J., Moin, P.: Reynolds stress and dissipation rate budgets in a turbulent channel flow. J. Fluid Mech. **194**, 5—44 (1988).

[28] Bagnold, R. A.: The physics of blown sand and desert dunes. London: Chapman and Hall (1941).

[29] Greeley, R., Iversen, I. D.: Wind as a geological process. Cambridge University Press. (1985).

[30] Lettau, K., Lettau, H. H.: Experimental and micrometeorological field studies of dune migration

       In: Exploring the world's driest climate. (eds. Lettau, H. H., Lettau, K.) University of Wisconsin, Madison, pp. 123—456 (1978).

[31] Press, W. R., Flannery, B. P., Teukolsky, S. A., Vetterling, W. T.: Numerical recipes, p. 381. Cambridge University Press (1986).

[32] Tsoar, H.: Profile analysis of sand dune and their steady state signification. Geografiska Annaler **67A**, 47—59 (1985).

[33] Cook, N. J.: Wind tunnel simulation of the adiabatic atmospheric boundary layer by roughness, barrier and mixing device methods. Building Res. Est. Paper 6, 157—176 (1977).

Authors' addresses: J. C. R. Hunt, Department of Applied Mathematics and Theoretical Physics, University of Cambridge, Silver Street, Cambridge, CB3 9 EW; W. S. Weng and D. J. Carruthers, CERC Lfd, 3 DV Kings Parade, Cambridge V CB2 1 SJ A. Warren and G. F. S. Wiggs, Department of Geography, University College London, 26 Bedford Way, London WC1H OAP; I. Livingstone, Department of Geography, Coventry Polytechnic, Priory Street, Coventry CV1 5FB; I. Castro, Department of Mechanical Engineering, University of Surrey, Guildford, Surrey GO2 5XH, United Kingdom.

Acta Mechanica (1991) [Suppl] 2: 23—35

On the temporal-spatial variation
of sediment size distributions

O. E. Barndorff-Nielsen and **M. Sørensen**, Aarhus, Denmark

Summary. A mathematical-physical model for the effect of erosion and deposition on the temporal and spatial variation of the size distributions of sediments is proposed and investigated. The model is closely related to the logarithmic hyperbolic distribution and preserves the hyperbolic shape of log-size distributions. In order to facilitate specification of the model in applications the concept of a 'local erosion time' is introduced. The local erosion time gives the local time scale of grain sorting changes. Explicit formulae for the temporal-spatial variation of the hyperbolic parameters predicted by the model are derived for stable bed forms and alluvial streams. The results are compared to field observations. Also the effect of the spatial gradient of the grain sorting is considered.

1 Introduction

A mathematical model for the effect of erosion and deposition on the sediment mass-size distribution at a fixed location was proposed in Barndorff-Nielsen and Christiansen [8]. A characteristic property of this model is that if, at some point in time, the size distribution is log-hyperbolic, then it will stay within this class of distributions at subsequent times. This property reflects the ubiquity of log-hyperbolic size distributions of sand samples from environments where the conditions determining the sediment composition are stable and homogeneous, cf. Bagnold [1], Bagnold and Barndorff-Nielsen [4], and Barndorff-Nielsen and Christiansen [8].

When studying sorting phenomena that take place in space as well as in time, it is often of interest to compare how far the changes have progressed at different points in space. In this paper we extend the model of Barndorff-Nielsen and Christiansen [8] to comprise both temporal and spatial variations. Also the extended model preserves the log-hyperbolic shape of the size distributions. A first attempt to obtain a model for the spatial variation of sediment size distributions in the particular case of a gently sloping sand formation was made in Barndorff-Nielsen [6]. The results of that paper, resting on an analogy of Taylor's 'frozen field hypothesis', is in accordance with the results obtained here by a rather different route.

In Section 2 a general model for the temporal-spatial variation of sediment size distributions is given together with necessary background material on hyperbolic distributions. A particular way of specifying this model is proposed in Section 3. In order to do this we introduce the new concept of a 'local erosion time' that gives the local time scale of the changes in the sediment grading.

The model in Section 3 gives simple expressions for the temporal and spatial variation of the hyperbolic parameters, which may be compared to detailed field measurements. In

Sections 4 and 5 we apply the model to stable bed forms and to alluvial streams, respectively, and we compare the results to available field observations. In particular, the model provides an explanation of the linear increase in the typical log grain size up the windward side of a small barchanoid dune reported in Barndorff-Nielsen et al. [10]. The formulae derived in Sections 3 and 4 facilitate future field tests of our model. In the final Section 6 two shortcomings of our model are discussed and solutions are proposed. One problem is that the model may not give valid results when applied over long time periods. The other problem considered is the effect of the spatial gradient of the grain sorting.

2 General theory

In Barndorff-Nielsen and Christiansen [8] a model for the temporal changes in the size distribution at one particular location is proposed. Here we shall extend their ideas to obtain a model that can predict spatial as well as temporal variations in the size distribution. In analogy with the Barndorff-Nielsen and Christiansen model, we assume that initially the size distribution is log-hyperbolic, and we consider erosion-deposition patterns that preserve this distribution type.

The probability density function of the hyperbolic distribution, as defined in Barndorff-Nielsen [5], is for fixed values of the four parameters μ, $\delta > 0$, and $0 \leq \beta < \alpha$ given by

$$p(s) = a(\delta, \alpha, \beta) \exp\left[-\alpha \sqrt{(\delta^2 + (s - \mu)^2)} + \beta(s - \mu)\right]. \tag{2.1}$$

Here

$$a(\delta, \alpha, \beta) = \sqrt{(1 - \varrho^2)} \Big/ \left[2K_1\big(\alpha\delta \sqrt{(1 - \varrho^2)}\big)\right], \tag{2.2}$$

K_1 being a Bessel function, and

$$\varrho = \beta/\alpha. \tag{2.3}$$

The parameters μ and δ specify the location and the scale, respectively, of the hyperbolic distribution. The parameter ϱ, which ranges from -1 to 1, is an important measure of the asymmetry of the distribution; cf. Barndorff-Nielsen et al. [9], where this is termed the tilt. For $\varrho = 0$ the distribution is symmetric around μ. We shall be concerned here with the following alternative measures of location and scale: The mode point

$$v = \mu + \delta\varrho(1 - \varrho^2)^{-1/2}, \tag{2.4}$$

and the square root of the curvature of the logarithmic probability density function at the mode point v, i.e.

$$\tau = (\alpha/\delta)^{1/2} (1 - \varrho^2)^{3/4}. \tag{2.5}$$

The parameter τ expresses the steepness of the probability density function near the mode point. The larger the value of τ the greater the steepness and the more the distribution is concentrated near the mode point. We shall refer to τ as the sorting.

As measures of kurtosis and skewness of the hyperbolic distribution we adopt

$$\xi - \left[1 + \alpha\delta \sqrt{(1 - \varrho^2)}\right]^{-1/2} \tag{2.6}$$

and

$$\chi = \varrho\xi. \tag{2.7}$$

The parameters ξ and χ are invariant under changes of location and scale. Their domain of variation is a triangle, the so-called hyperbolic shape triangle. (It should be noted that ξ and χ are not exactly equal to the kurtosis and the skewness of the hyperbolic distribution as traditionally defined in terms of moments.) For further discussion of these invariant parameters and the hyperbolic shape triangle, see Barndorff-Nielsen et al. [7], Barndorff-Nielsen and Christiansen [8] and Barndorff-Nielsen et al. [9].

In Barndorff-Nielsen and Christiansen's [8] model the changes of the log-hyperbolic size distribution due to erosion and deposition are specified by two quantities ε and $\varkappa$. Net erosion corresponds to a positive ε and a negative $\varkappa$, and vice versa for net deposition (ε and $\varkappa$ having the same sign does not make physical sense). Under that model the location parameter μ and the scale parameter δ are constant whereas ε and $\varkappa$ affect both ξ and ϱ and hence also χ, v and τ. Further, the relation between ξ and ϱ is given by

$$\xi = \left\{1 + ce^{-(\varkappa/\varepsilon)\varrho}\sqrt{(1-\varrho^2)}\right\}^{-1/2} \tag{2.8}$$

where c is a constant. Note that ε and $\varkappa$ enter (2.8) only through the ratio $\lambda = -(\varkappa/\varepsilon)$. Figure 1 shows some typical examples of the curve (2.8), for various values of λ and c. The interpretation of these curves is that a population of sand grains whose distributional shape is represented by a single point in the hyperbolic shape triangle will, when subjected to erosion or deposition, be continuously transformed into new log hyperbolic populations having the same μ and δ values whereas ξ and ϱ (and χ) change in such a way that the shape point moves along the particular curve (2.8) determined by the initial position and the ratio λ.

In the present paper we consider the situation where the hyperbolic parameters depend not only on time but also on a spatial coordinate x, which may be 1-, 2- or 3-dimensional. The size distribution at location x at time t is supposed to be log-hyperbolic with parame-

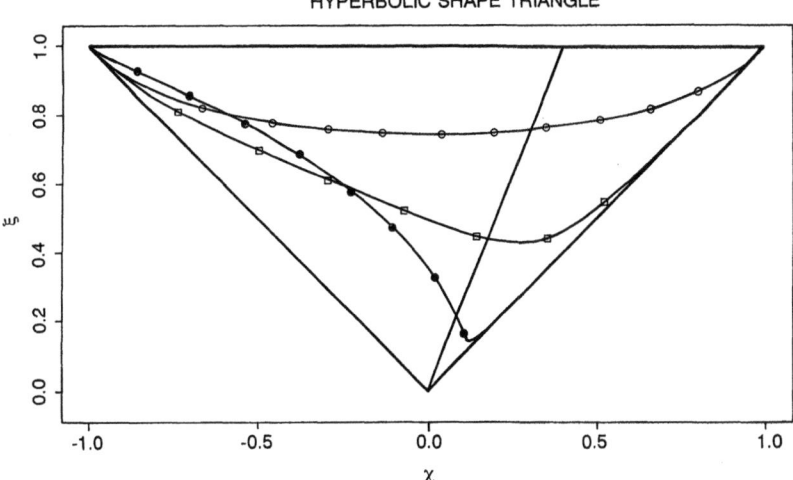

Fig. 1. The hyperbolic shape triangle with four examples of the erosion-deposition curves, as determined by equation (2.8) and by the relations $\chi = \varrho\xi$ and $\lambda = -\varkappa/\varepsilon$. The straight curve corresponds to a fixed value of ϱ whereas the other three are specified by $\bigcirc$ $\lambda = 0$, $c = 0.8$ $\square$ $\lambda = 1$, $c = 3$ $\bullet$ $\lambda = 3$, $c = 7$

ters $\mu(x, t)$, $\delta(x, t)$, $\alpha(x, t)$ and $\beta(x, t)$. If we allow ε and $\varkappa$ to depend on x and t, the dynamical equations of Barndorff-Nielsen and Christiansen [8] generalize to

$$\dot{\alpha}(x, t) = -\varkappa(x, t) \tag{2.9}$$

$$\dot{\beta}(x, t) = \varepsilon(x, t) - \varkappa(x, t)\, \varrho(x, t) \tag{2.10}$$

$$\dot{\delta}(x, t) = 0 \tag{2.11}$$

$$\dot{\mu}(x, t) = 0. \tag{2.12}$$

Here a dot denotes the partial derivative with respect to t. Combining (2.9) and (2.10) we find that

$$\lambda(x, t)\, \dot{\varrho}(x, t) = \dot{\alpha}(x, t)/\alpha(x, t) \tag{2.13}$$

where

$$\lambda(x, t) = -\frac{\varkappa(x, t)}{\varepsilon(x, t)}. \tag{2.14}$$

Note that $\lambda(x, t) \geqq 0$. The equation (2.13) is easily solved if $\lambda(x, t)$ does not depend on t. We shall consider the somewhat more general situation where λ is of the form

$$\lambda(x, t) = k(x)\, h\big(\varrho(x, t)\big) \tag{2.15}$$

for some functions k and h. Let H be a primitive of h, i.e. $H' = h$. Then (2.13) can be re-written as

$$k(x)\, \frac{\partial}{\partial t}\, H\big(\varrho(x, t)\big) = \frac{\partial}{\partial t}\, \ln\big(\alpha(x, t)\big), \tag{2.16}$$

so that

$$k(x)\, \big[H\big(\varrho(x, t)\big) - H\big(\varrho(x, 0)\big)\big] = \ln\left[\alpha(x, t)/\alpha(x, 0)\right]. \tag{2.17}$$

It will usually be possible to determine the variation of α from the simple equation (2.9), and provided H is invertible (i.e. if $\lambda > 0$), the variation of ϱ follows via (2.17). The variation of the other hyperbolic parameters mentioned can be found by substituting $\alpha(x, t)$ and $\varrho(x, t)$ in (2.3)−(2.7), using that μ and δ are constant in time (cf. (2.11) and (2.12)). In particular, insertion of (2.17) in (2.6) gives a relation between the tilt parameter $\varrho(x, t)$ and the kurtosis parameter $\xi(x, t)$:

$$\xi(x, t) = \left[1 + c(x) \exp\big\{k(x)\, H\big(\varrho(x, t)\big)\big\}\, \sqrt{\{1 - \varrho^2(x, t)\}}\right]^{-1/2} \tag{2.18}$$

with

$$c(x) = \alpha(x, 0)\, \delta(x)\, e^{-k(x)H(\varrho(x, 0))}.$$

As t varies with x fixed the parameters (χ, ξ) thus move in the hyperbolic shape triangle along a curve defined by k, H and the initial conditions (at time $t = 0$).

Let us conclude this section by briefly considering the case where $\lambda(x, t)$ does not depend on t. This means that the function h can, without loss of generality, be set equal to one, cf. formula (2.15). Hence $H(\varrho) = \varrho$, so that

$$\varrho(x, t) = \varrho(x, 0) + \lambda(x)^{-1} \ln\{\alpha(x, t)/\alpha(x, 0)\} \tag{2.19}$$

and

$$\xi(x, t) = \left[1 + c(x) \exp\{\lambda(x)\, \varrho(x, t)\}\, \sqrt{\{1 - \varrho^2(x, t)\}}\right]^{-1/2}. \tag{2.20}$$

The curve (2.20) equals the one derived in Barndorff-Nielsen and Christiansen [8], cf. (2.8), even though we allow $\varkappa$ and ε to depend on time and only require that their ratio is constant in time. This conclusion, of course, follows from the fact that the relation between ϱ and α depends on $\varkappa$ and ε only through their ratio λ, cf. (2.13).

3 The local erosion time

Motivation and definition

So far the variation of the quantities ε and $\varkappa$ in the sorting model of Section 2 has received only very limited empirical investigation, see Barndorff-Nielsen and Christiansen [8]. Therefore there is, at the moment, not a solid empirical basis for choosing a specification of the spatial and temporal variation of ε and $\varkappa$. We can get around this problem by introducing the concept of a local erosion time, a concept that might also turn out to be useful in other contexts.

The *local erosion time* is a function $\mathcal{T}(x, t)$ of the spatial coordinate x and of the astronomical time t, which gives the time scale of the changes in the size distribution at location x and time t in the following sense. We assume that the size distribution at each location goes through the sequence of stages given by the model in Section 2 with values of ε and $\varkappa$ that are constant in time, but may differ between locations. Moreover, we allow the speed, at which the size distribution progresses through the sequence of stages defined by $\varepsilon(x)$ and $\varkappa(x)$, to vary from one position to another. At position x the size distribution has at time t reached the stage indexed by the local erosion time $\mathcal{T}(x, t)$. A reasonable assumption is that the speed at which the size distribution at a particular location changes is proportional to the degree of sediment activity at that location. We quantify the degree of sediment activity by the rate $q(x, t)$ of sediment transport at location x and time t. In view of these considerations we define the local erosion time as

$$\mathcal{T}(x, t) = \int\limits_0^t [q(x, s)/\bar{q}(x_0)]\, ds, \tag{3.1}$$

where x_0 is a suitably chosen reference point and $\bar{q}(x_0)$ is the typical transport rate at x_0. The quantity $\bar{q}(x_0)$ can be the average transport rate at x_0 or $q(x_0, t_0)$ for some suitably chosen fixed time t_0. Note that the total transport from time 0 to time t at position x equals $\bar{q}(x_0)\, \mathcal{T}(x, t)$.

In the stationary case, where q does not depend on time, formula (3.1) simplifies to

$$\mathcal{T}(x, t) = tq(x)/q(x_0). \tag{3.2}$$

In this important situation the local erosion time at the reference point x_0 equals the astronomical time.

Of course, we could have focused on another quantity than the transport rate in defining the degree of sediment activity. In our context of sorting changes the transport rate seems most relevant, but in studies of, for instance, aeolian abrasion the kinetic energy flux would be more to the point, see Greeley and Iversen [14].

A temporal-spatial model for erosion and deposition

When $\varkappa$ does not depend on time, we find from (2.9) that

$$\alpha(x, s) = \alpha(x, 0) - \varkappa(x)\, s. \tag{3.3}$$

Proceeding as discussed in the previous subsection, the value of α at location x at time t is found by substituting the local erosion time $\mathcal{T}(x, t)$ for s in (3.3), i.e.

$$\alpha(x, t) = \alpha(x, 0) - \varkappa(x)\,\mathcal{T}(x, t). \tag{3.4}$$

The value of ϱ, found from (2.19), is

$$\varrho(x, t) = \varrho(x, 0) + \lambda(x)^{-1} \ln\left[1 - \big(\varkappa(x)/\alpha(x, 0)\big)\,\mathcal{T}(x, t)\right], \tag{3.5}$$

where $\lambda(x) = -\varkappa(x)/\varepsilon(x)$. The variation of the other hyperbolic parameters is determined by (3.4) and (3.5) as we shall discuss later.

We can obtain the same results if, in the general model of Section 2, we set

$$\varkappa(x, t) = \varkappa^*(x)\,q(x, t) \tag{3.6}$$

with $\varkappa^*(x) = \varkappa(x)/\bar{q}(x_0)$, and

$$\varepsilon(x, t) = \varepsilon^*(x)\,q(x, t) \tag{3.7}$$

with $\varepsilon^*(x) = \varepsilon(x)/\bar{q}(x_0)$. Thus we have found, at least partly, a way of specifying $\varkappa(x, t)$ and $\varepsilon(x, t)$ in the general model in terms of the transport rate — a much more well-studied quantity than ε and $\varkappa$. As a first attempt and in the absence of indications to the contrary one could assume that the sequences of events are the same at all locations, i.e. one could set $\varkappa(x)$ and $\varepsilon(x)$ equal to constants. This will indeed be done in the next two sections.

By substitution in (2.4), (2.5) and (2.6) of (3.4) and (3.5) we find that

$$v(x, t) = \mu(x) + \delta(x)\,\varrho(x, t)\,[1 - \varrho(x, t)^2]^{1/2}, \tag{3.8}$$

$$\tau(x, t) = \delta(x)^{-1/2}\,[\alpha(x, 0) - \varkappa(x)\,\mathcal{T}(x, t)]^{1/2}\,[1 - \varrho(x, t)^2]^{3/4} \tag{3.9}$$

and

$$\xi(x, t) = \left[1 + \delta(x)\,\{\alpha(x, 0) - \varkappa(x)\,\mathcal{T}(x, t)\}\,\sqrt{\{1 - \varrho(x, t)^2\}}\,\right]^{1/2}. \tag{3.10}$$

Note that for small erosion times or, more precisely, for

$$\varkappa(x)\,\mathcal{T}(x, t)/\alpha(x, 0) \ll 1 \tag{3.11}$$

we have

$$\varrho(x, t) = \varrho(x, 0) + [\varepsilon(x)/\alpha(x, 0)]\,\mathcal{T}(x, t). \tag{3.12}$$

Condition (3.11) states that the change in α, given by $\varkappa(x)\,\mathcal{T}(x, t)$, is small compared to the magnitude of α. If, moreover, $|\varrho(x, t)| \ll 1$, we find that

$$v(x, t) = \mu(x) + \delta(x)\,\varrho(x, 0) + [\varepsilon(x)\,\delta(x)/\alpha(x, 0)]\,\mathcal{T}(x, t), \tag{3.13}$$

and

$$\tau(x, t) = \big(\alpha(x, 0)/\delta(x)\big)^{1/2}\left[1 - \frac{1}{2}\,\varkappa(x)\,\mathcal{T}(x, t)/\alpha(x, 0)\right]. \tag{3.14}$$

We see that, under the conditions imposed, the mode point v and the sorting τ vary linearly with the local erosion time.

4 Stable bed forms

We shall now relate the local erosion time and the temporal-spatial model for erosion and deposition of Section 3 to the rate of sand transport across a stable bed form. Suppose, for concreteness, that the bed form is a dune of a fixed shape moving with a constant velocity v in the direction of the fluid flow. It is not important here whether the dune is aeolian or alluvial. It might also be a smaller bed form than a dune. We assume that the dune is two-dimensional and that its profile at time t is given by

$$y(x, t) = h(x - vt). \tag{4.1}$$

Here x is a coordinate in the fluid flow direction. The function $h(x)$ determines the surface of the dune at time $t = 0$. In a coordinate system that follows the dune the x-coordinate is given by $x' = x - vt$, and in this coordinate system $h(x')$ obviously determines the dune at any time.

It is well-known, see e.g. Exner [12] or Bagnold [1], (p. 200) for two independent derivations, that a simple continuity argument implies that the transport rate $q(x')$ at location $x' = x - vt$ is related to $h(x')$ by

$$q(x') = vkh(x') + q(0). \tag{4.2}$$

Here k denotes the bulk density of the sand, while $x' = 0$ defines a reference point relative to the dune where we have taken $h(0) = 0$. The equation (4.2) predicts that the difference $q(x') - q(0)$ of the local transport rates is proportional to the local height of the dune above a fixed reference plane. The relation (4.2) can not be assumed to hold on a slip face.

Since we have assumed the dune to be stationary in a coordinate system that follows the dune, the local erosion time can be found from (3.2). Thus at location $x' = x - vt$

$$\mathcal{T}(x', t) = t\{1 + vkh(x')/q(0)\}, \tag{4.3}$$

where we have chosen $x' = 0$ as the reference point at which the local erosion time equals astronomical time.

Substitution of (4.3) into (3.4), (3.5) and (3.8)–(3.10) gives expressions for the variation of the hyperbolic parameters over the dune at a fixed time t. Here we will discuss only the situation where, at time $t = 0$, the size distribution at all positions on the dune is the same log-hyperbolic distribution with parameters α, β, δ, μ and $\varrho = \alpha/\beta$. Moreover, we assume that $\varkappa$ and ε are constant. Then, at time t, the parameter α increases linearly with the height of the dune. Specifically, at location x',

$$\alpha(x', t) = \alpha - \varkappa t - [\varkappa vkt/q(0)] h(x'). \tag{4.4}$$

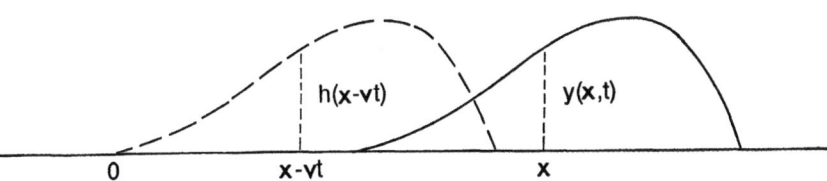

Fig. 2. Illustrating the idealized movement of a stable bed form moving with constant velocity v, $y(x, t) = h(x - vt)$ being the height of the bed form at position x at time t

The stoss side of the dune is an erosional regime, so $\varkappa < 0$ and $\varepsilon > 0$. The rate by which $\alpha(x', t)$ increases with the height $h(x')$ increases proportionally to the time t.

The other hyperbolic parameters vary as more complicated functions of the height. However, it is easily seen that the tilt parameter $\varrho(x', t)$ and the typical logarithmic grain size $v(x', t)$ increase with the height $h(x')$. Some elementary calculations reveal that the sorting parameter $\tau(x', t)$ increases with height as long as $\varrho(x', t)$ is smaller than $\varrho_0 = \sqrt{(1 + \omega^2)} - \omega$, where $\omega = -3\varepsilon/(2\varkappa)$. Note that ϱ_0 is a positive number varying between 0 ($\omega = \infty$) and 1 ($\omega = 0$). For $\varepsilon/\varkappa = -1$ we find that $\varrho_0 = 0.30$. As soon as $\varrho(x', t)$ grows larger than ϱ_0, τ starts to decrease with height. Similarly, the kurtosis parameter $\xi(x', t)$ decreases with height for $\varrho(x', t)$ smaller than $\varrho_1 = \sqrt{(1 + \bar{\omega}^2)} - \bar{\omega}$, $\bar{\omega} = -\varepsilon/(2\varkappa)$, and increases with height when $\varrho(x', t)$ is larger than ϱ_1. Also ϱ_1 is a number between 0 and 1, and $\varrho_1 > \varrho_0$. For $\varepsilon/\varkappa = -1$ we find $\varrho_1 = 0.62$. Barndorff-Nielsen and Christiansen [8] found that the value $\varepsilon/\varkappa = -1$ fitted data from a microtidal flat well. Figure 5 in the same paper indicates that $\varepsilon/\varkappa$ might be smaller for aeolian environments. Indications in the same direction can be found in Section 5.4.3 of Hartmann [15]. Very often the tilt of the size distribution of aeolian sand dunes is negative or, if positive, not far above zero. Therefore, typically $\tau(x', t)$ increases and $\xi(x', t)$ decreases with height.

If the variation in α is small compared to its magnitude and if the size distributions are only moderately skewed we can use the simpler formulae $(3.12) - (3.14)$. The precise conditions are $\varkappa t/\alpha \ll 1$ and $|\varrho| \ll 1$. If these are satisfied, we find that

$$\varrho(x', t) = \varrho + (\varepsilon/\alpha)\, t + \left[\varepsilon v k t / (\alpha q(0)) \right] h(x'), \tag{4.5}$$

$$v(x', t) = \mu + \delta\varrho + (\varepsilon\delta/\alpha)\, t + \left[\varepsilon\delta v k t / (\alpha q(0)) \right] h(x') \tag{4.6}$$

and

$$\tau(x', t) = (\alpha/\delta)^{1/2} \left[1 - \frac{1}{2} (\varkappa/\alpha)\, t - \frac{1}{2} \left\{ \varkappa v k t / (\alpha q(0)) \right\} h(x') \right]. \tag{4.7}$$

As in (4.4), we see that at a fixed time t the asymmetry, as measured by ϱ, the mode point v and the sorting τ increase linearly with the height. As for α the rate of increase with height increases proportionally to time t.

Finally, we consider briefly a dune with a linear stoss side, i.e. with $h(x') = cx'$. Then, obviously, $\alpha(x', t)$ increases in general linearly with x' on the stoss side. Under the approximations $(4.5) - (4.7)$, the same is true of $\varrho(x', t)$, $v(x', t)$ and $\tau(x', t)$.

It should be noted that the sorting process caused by the sand transport on the stoss side of the dune only takes place in a relatively thin sand layer. Therefore inhomogeneities in the sorting of the sand in the interior of the dune might cause deviations from the predictions of the model.

Qualitatively, the predictions of coarser typical grain size and better sorting at the dune crest than at the stoss are in accordance with observations from linear dunes due to Folk [13] and Lancaster [16]. The study of the variation in particle size distribution over a small barchanoid dune reported in Barndorff-Nielsen et al. [10] revealed a linear increase of the typical log grain size v with distance downwind on the stoss up to the crest. This accords well with (4.6) because the profile of the stoss was close to linear.

5 Longitudinal grain sorting in alluvial streams

In this section the longitudinal variation of the size distribution of the bed sediment in a river will be discussed in the light of the concept of local erosion time introduced in Section 3. We choose a starting point similar to that of Deigaard and Fredsøe [11] in their model of longitudinal grain sorting in rivers. Our methods, however, are entirely different from theirs and from the point of view of physics much more primitive.

The longitudinal profile of the river bed is supposed to be described by a function $h(x)$, where x is a coordinate increasing in the flow direction. Initially the bed material is supposed to have the same size distribution in the whole length of the river. The water discharge per unit width Q is assumed constant along the river. In natural rivers the water discharge per unit width usually increases slightly in the flow direction, but Deigaard and Fredsøe [11] argue that the effect of this can be disregarded when studying grain sorting. In our approach it would not be difficult to treat the case of a longitudinally varying Q. We suppose the river to be in a steady state defined by $h(x)$ and the water discharge.

It seems likely that the time scale of the grain sorting process is determined mainly by the bedload transport rate so that we need not take into account the transport rate of suspended material. We therefore use the bedload transport rate in the definition (3.2) of the local erosion time $\mathscr{T}(x, t)$. Bagnold [2] found that the bedload transport rate q_b is related to the stream power per unit width given by

$$\omega = \gamma Q I. \tag{5.1}$$

Here γ is the specific gravity of the water, and I is the energy gradient

$$I(x) = -h'(x). \tag{5.2}$$

In Bagnold [3] it was found that the empirical expression

$$q_b = c(\omega - \omega^*)^{3/2} D^{-2/3} s^{-1/2}, \tag{5.3}$$

for some constant c, closely correlates flume data on measured rates of unsuspended bedload transport with data on the like transport rates in a wide variety of natural rivers. In (5.3) ω^* is the threshold value of the stream power ω at which sediment begins to be moved, D denotes the flow depth, and s is the typical (modal) grain size.

As in the rest of the paper we assume that the bed sediment has a log-hyperbolic size distribution. Therefore, $s(x, t) = \exp\big(v(x, t)\big)$, so actually the transport rate at a given location x changes slowly with time. Hence the situation is not entirely stationary, and the most precise procedure would be to find the erosion time by combining (3.1), (3.5) and (3.8). Here we avoid this complication by considering only time points t small enough that the factor $\exp\big[1/2\,\big(v(x_0, 0) - v(x, t)\big)\big]$ can safely be ignored in the definition of the local time. Here x_0 is the reference point chosen. The size-dependent factor can certainly be set equal to one, under the conditions ensuring (3.13), i.e. $1/2\,\varepsilon\delta\alpha^{-1}\,\mathscr{T}(x, s) \ll 1$ with $\mathscr{T}(x, s)$ defined by (5.4) below. Also ω^* depends on the size distribution, but again we will only consider situations where ω^* can be assumed constant.

Under our assumptions the local erosion time at location x is given by

$$\mathscr{T}(x, t) = t \left\{\frac{I(x) - I^*}{I(x_0) - I^*}\right\}^{3/2} \left\{\frac{D(x_0)}{D(x)}\right\}^{2/3}, \tag{5.4}$$

where $I^* = \omega^*/\gamma Q$ is the smallest gradient at which sediment transport takes place. As usual x_0 is a reference point. Since, in natural rivers, $I(x)$ decreases and $D(x)$ increases downstream, we see that the local erosion time decreases in the flow direction. The spatial variation of the size distribution can now be found by substituting (5.4) in the equations (3.4), (3.5) and (3.8)—(3.10). We assume that ε and $\varkappa$ are constants. The upper part of a river is an erosional regime, so $\varkappa < 0$ and $\varepsilon > 0$. Of course, this is a rough model which ignores local and seasonal variations.

We find that $\alpha(x, t)$, the asymmetry parameter $\varrho(x, t)$ and the typical grain size $\exp\big(v(x, t)\big)$ all decrease downstream. The sorting $\tau(x, t)$ decreases with x provided $\varrho(x, t) < \varrho_0$, defined in section 4. Otherwise τ increases downstream. Similarly, the kurtosis parameter $\xi(x, t)$ increases in the flow direction when $\varrho(x, t) < \varrho_1$, where ϱ_1 is defined in Section 4, and decreases otherwise. The fact that the typical grain size decreases downstream implies that after a long time the correctly defined local erosion time, including the size-dependent term, decreases downstream more slowly than $\mathscr{T}(x, t)$ defined by (5.4). At first the gradient is reduced by the factor $\exp\big(-1/2\ \varepsilon\delta\alpha^{-1}\mathscr{T}(x, t)\big)$.

In rivers it is generally found that the typical grain size decreases in the downstream direction, see Deigaard and Fredsøe [11]. This is in accordance with our prediction. The same result follows from the models for longitudinal grain sorting in alluvial streams due to Rana et al. [17] and Deigaard and Fredsøe [11]. The latter model also predicts that if initially the bed material has the same log-normal size distribution in the entire length of the river then the size distribution will later tend to be increasingly non-normal, in fact ultimately hyperbolic, in the downstream direction. The same trend is found in our model if the initial size distribution is close to log normal ($\xi = 0$). Deigaard and Fredsøe [11] also predict that the skewness becomes more negative in the flow direction, again in accordance with our results.

6 Discussion

The mathematical model for the spatial variation of size distributions presented in Sections 2 and 3 is obviously a very simple one. An attractive consequence of this is that the model gives simple expressions for the predicted temporal-spatial variation of the hyperbolic parameters. Therefore, the model can be tested by detailed field measurements. We have already compared our theoretical results for bed forms and rivers to existing field observations. We hope that more comparisons of this kind will be done in the future.

On the other hand our simple model should be used thoughtfully. In the following we discuss in some detail two problems which might well be encountered in applications, and we propose solutions to them.

Sorting changes over long time spans

The model of Section 2 can, in general, not be expected to work in long time periods when $\lambda(x, t)$, defined by (2.14), does not depend on time. Consider, for instance, the case where $\varkappa(x)$ is independent of time. First suppose that $\varkappa(x) > 0$, i.e. that the regime is depositional. From formula (3.3) it follows that for $t > \alpha(x, 0)/\varkappa(x)$ the value of $\alpha(x, t)$ is negative, in disagreement with the range of variation of the hyperbolic parameter α ($\alpha > 0$). Also in an erosional regime there are problems. Here $\alpha(x, t) \to \infty$ as $t \to \infty$, so by (2.19) the same is true of $\varrho(x, t)$, whereas the range of variation of ϱ is $|\varrho| < 1$.

We can get around these problems, however, by allowing $\lambda(x, t)$ to depend on $\varrho(x, t)$ in the way discussed below. The possibility that $\lambda(x, t)$ depends on $\varrho(x, t)$ is already built into the general model in Section 2. In formula (2.15) we set

$$h(\varrho) = (1 - \varrho^2)^{-\psi} \tag{6.1}$$

for some $\psi \geq 0$. A physical interpretation of this specification of $\lambda(x, t)$ is as follows. When the parameter ϱ is numerically close to one, the size distribution of the bed material is highly asymmetric so that the influence (given by $\varkappa$) on the sorting process caused by the differential sizes of the bed grains dominates over the ε-erosion or ε-deposition. A consequence of (6.1) is that $\dot{\varrho}(x, t)$ tends to zero as $|\varrho(x, t)|$ approaches one, cf. (2.13).

A primitive of h can be found explicitly when ψ is a multiple of one half. For $\psi = 1$ we find the primitive

$$H(\varrho) = \frac{1}{2} \ln\left[(1 + \varrho)/(1 - \varrho)\right]. \tag{6.2}$$

With this choice of ψ it follows from (2.17) that $\varrho(x, t) \to 1$ as $\alpha(x, t) \to \infty$ and that $\varrho(x, t) \to -1$ as a $\alpha(x, t) \to 0$. This is because $H(\varrho) \to \pm\infty$ as $\varrho \to \pm 1$. Note also that $H(\varrho) = \varrho + O(\varrho^3)$, implying that, not too far from $\chi = 0$, the curve in the hyperbolic shape triangle, defined by (2.18) with H given by (6.2), is very much like the one for $\lambda(x, t)$ independent of t, cf. (2.20). Thus when $|\varrho|$ is not too large, we can safely ignore the complication (6.1) and use the model with $\lambda(x, t)$ depending on x only.

For values of ψ smaller than one H stays bounded as $|\varrho| \to 1$, which in view of (2.17) is not satisfactory. As ψ increases $H(\varrho)$ becomes more complicated and deviates more from the identity line. For instance for $\psi = 3$

$$H(\varrho) = \frac{\varrho(5 - 3\varrho^2)}{8(1 - \varrho^2)^2} + \frac{3}{16} \ln\left[\frac{1 + \varrho}{1 - \varrho}\right].$$

The modification of $\lambda(x, t)$ given by (6.1) solves the problem concerning the magnitude of ϱ. We are still faced with the problem that, under depositional conditions, $\alpha(x, t)$ may become negative. A simple solution would be to let $\varkappa(x, t)$ go to zero as $\alpha(x, t)$ approaches zero. The physical explanation would be that a size distribution becomes more stable as α approaches 0 and ϱ goes towards -1, meaning that the distribution becomes very broad with mainly fine grains. A simple way of specifying behaviour of this type is to set

$$\varkappa(x, t) = \bar{\varkappa}(x, t)\, \alpha(x, t), \tag{6.3}$$

where $\bar{\varkappa}(x, t)$ is a given bounded function of x and t that is independent of α. For instance, $\bar{\varkappa}$ could, in analogy with (3.6), be taken proportional to the transport rate $q(x, t)$. The specification (6.3) implies that

$$\alpha(x, t) = \alpha(x, 0) \exp\left[-\int_0^t \bar{\varkappa}(x, s)\, ds\right], \tag{6.4}$$

which is always positive.

The effect of the spatial sorting gradient

A quite different, but presumably more important, objection to the model presented in Sections 2 and 3 is that the modelling of the sorting process is purely local. Long range spatial interactions, which are likely to develop over long time spans, can certainly not be accomodated to the model. Even the effect of the local spatial gradient in the grain sorting can not be built into the model as it stands. In order to get an idea about the possible effect of the local spatial gradient we consider the following generalization of our model.

Barndorff-Nielsen and Christiansen [8] derived the equations (2.9)−(2.12) by assuming that in the time interval $(t, t + dt)$ the density $p(s)$ of the size distribution (s denotes the logarithm of the grain diameter) is modified by the multiplication by two factors. The first factor

$$\exp(\varepsilon s dt) \tag{6.5}$$

corresponds to the general ε-erosion/deposition. The second factor expresses the effect of the differential sizes of the bed grains and is given by

$$\exp\left[\varkappa\left\{\sqrt{(\delta^2 + (s - \mu)^2)} - \varrho(x, t)(s - \mu)\right\} dt\right], \tag{6.6}$$

where μ, δ and $\varrho(x, t)$ are parameters of the hyperbolic distribution of the bed sediment at location x at time t.

A natural way of introducing the effect of the spatial grading gradient is to multiply onto $p(s)$ a third factor dependent on the size distribution at the typical origin $\tilde{x}$ of grains deposited at location x. A third factor preserving the log-hyperbolic distribution is

$$\exp\left[\varkappa'\left\{\sqrt{(\delta^2 + (s - \mu)^2)} - \varrho(\tilde{x}, t)(s - \mu)\right\} dt\right], \tag{6.7}$$

where $\varkappa'$ is negative irrespective of whether the regime is erosional or depositional. It is important here that we assume the parameters μ and δ constant in space as well as in time. Let us for simplicity of presentation assume that x is a 1-dimensional coordinate increasing in the wind direction. The combined effect of the three factors (6.5)−(6.7) can be expressed in the differential equations

$$\frac{\partial}{\partial t} \alpha(x, t) = -\varkappa^+ \tag{6.8}$$

$$\frac{\partial}{\partial t} \beta(x, t) = \varepsilon + \gamma \frac{\partial}{\partial x} \varrho(x, t) - \varkappa^+ \varrho(x, t), \tag{6.9}$$

where $\varkappa^+ = \varkappa + \varkappa'$ and $\gamma = \varkappa'(\tau - \tilde{x}) < 0$. We have approximated $\varrho(x) - \varrho(\tilde{x})$ by $(x - \tilde{x}) \times \frac{\partial}{\partial x} \varrho(x, t)$. Presumably $|\varkappa'| < |\varkappa|$ so that $\varkappa^+$ has the same sign as $\varkappa$, i.e. negative in erosional regimes and positive under depositional conditions. Combining (6.8) and (6.9) we find that

$$\frac{\partial}{\partial t} \varrho(x, t) = \left\{\lambda^{-1} - \lambda' \frac{\partial}{\partial x} \varrho(x, t)\right\} \frac{\partial}{\partial t} \ln(\alpha(x, t)) \tag{6.10}$$

with $\lambda = -\varkappa^+/\varepsilon > 0$ and $\lambda' = \gamma/\varkappa^+$. The quantities ε, $\varkappa$ and γ could be allowed to depend on x and t.

Equation (6.10) is easily solved provided that λ and λ' are constant and that $\alpha(t)$ does not depend on x. Under these conditions, where the only spatially varying effect is that of the gradient,

$$\varrho(x, t) = \lambda^{-1} \ln(\alpha(t)/\alpha(0)) + \varrho(x - \lambda' \ln(\alpha(t)/\alpha(0)), 0) \tag{6.11}$$

solves (6.10). A similar result is found if either $\varkappa^+(x, t) = \alpha(x, 0) \bar{\varkappa}(t)$ or $\varkappa^+(x, t) = \alpha(x, t) \times \bar{\varkappa}(t)$, where $\bar{\varkappa}(t)$ is independent of x. In either case α does depend on x, while $\alpha(x, t)/\alpha(x, 0)$

does not. We see, by comparing (6.11) to (2.19), that the effect of the sorting gradient is that the initial spatial variation of the asymmetry parameter $\varrho(x, 0)$ is being shifted in the wind direction. To see this remember that under erosional conditions $\alpha(t)$ increases with t and $\lambda' > 0$. For depositional regimes it is the other way round: $\alpha(t)$ decreases and $\lambda' < 0$. This effect of the sorting gradient is in accordance with what one would expect.

References

[1] Bagnold, R. A.: The physics of blown sand and desert dunes. London: Methuen, (reprint published in 1973 by Chapman and Hall, London) 1941.

[2] Bagnold, R. A.: An approach to the sediment transport problem from general physics. U.S. Geol. Surv. Prof. Pap. **422**-I: 1—37 (1966).

[3] Bagnold, R. A.: An empirical correlation of bedload transport rates in flumes and natural rivers. Proc. R. Soc. Lond. A **372**, 453—473 (1980).

[4] Bagnold, R. A., Barndorff-Nielsen, O. E.: The pattern of natural size distributions. Sedimentology **27**, 199—207 (1980).

[5] Barndorff-Nielsen, O. E.: Exponentially decreasing distributions for the logarithm of particle size. Proc. R. Soc. Lond. A **353**, 401—419 (1977).

[6] Barndorff-Nielsen, O. E.: Aeolian erosion on a gently sloping sand formation. A note. Research Report 175. Dept. Theor. Statistics. Aarhus University (1989).

[7] Barndorff-Nielsen, O. E., Blæsild, P., Jensen, J. L., Sørensen, M.: The fascination of sand. In: A celebration of statistics (Atkinson, A. C., Fienberg, S. E., eds.) Centenary Volume of the International Statistical Institute, pp. 57—87. Springer, New York, 1985.

[8] Barndorff-Nielsen, O. E., Christiansen, C.: Erosion, deposition and size distributions of sand. Proc. R. Soc. Lond. A **417**, 335—352 (1988).

[9] Barndorff-Nielsen, O. E., Christiansen, C., Hartmann, D.: Distributional shape triangles with some applications in sedimentology. Acta Mechanica [Suppl. 2]: 37—47 (1991).

[10] Barndorff-Nielsen, O. E., Dalsgaard, K., Halgreen, C., Kuhlman, H., Møller, J. T., Schou, G.: Variation in particle size distribution over a small dune. Sedimentology **29**, 53—65 (1982).

[11] Deigaard, R., Fredsøe, J.: Longitudinal grain sorting by current in alluvial streams. Nord. Hydrol. **9**, 7—16 (1978).

[12] Exner, F. M.: Sitzber. Akad. Wiss. (Wien) **3—4**, 165 (1925).

[13] Folk, R. L.: Longitudinal dunes of the northwestern edge of the Simpson Desert, Northern Territory, Australia, 1. Geomorphology and grain size relationships. Sedimentology **16**, 5—54 (1971).

[14] Greeley, R., Iversen, J. D.: Wind as a geological process on Earth, Mars, Venus and Titan. Cambridge: Cambridge University Press, 1985.

[15] Hartmann, D.: Coastal sands of the Southern and Central part of the Mediterranean Coast of Israel — reflection of dynamic sorting processes. Ph. D.-Thesis, Faculty of Science, Aarhus University, 1988.

[16] Lancaster, N.: Grain-size characteristics of linear dunes in the Southwestern Kalahari. J. Sed. Petrol. **56**, 395—400 (1986).

[17] Rana, S. A., Simons, D. B., Mahmood, K.: Analysis of sediment sorting in alluvial channels. J. Hyd. Div. ASCE **99**, 1967—1980 (1973).

Authors' address: Prof. O. E. Barndorff-Nielsen and M. Sørensen, Department of Theoretical Statistics, Institute of Mathematics, Aarhus University, Ny Munkegade, DK-8000 Aarhus C, Denmark

Acta Mechanica (1991) [Suppl] 2: 37—47

Distributional shape triangles with some applications in sedimentology

O. E. Barndorff-Nielsen and C. Christiansen, Aarhus, Denmark, and D. Hartmann, Haifa, Israel

Summary. The location-scale invariant parameters χ and ξ of the hyperbolic distribution have a triangular domain of variation which is referred to as the hyperbolic shape triangle. There are close analogies between the hyperbolic distributions, the generalized logistic distributions and the beta distributions, which make it possible also to define a logistic or a beta shape triangle.

Using a population concept of sediment samples it is shown that both the hyperbolic shape triangle and the beta shape triangle provide useful information on variations of grain size distributions, in cross-shore transects. This information can be related to the process-oriented erosion/deposition model developed previously in connection with the hyperbolic shape triangle.

1 Introduction

In the subject area of sorting of sand by wind or water the hyperbolic shape triangle (Barndorff-Nielsen et al. [4]) has turned out to be useful, as a 'blackboard', for relating characteristics of grain size distributions to the dynamics of the sorting processes (Barndorff-Nielsen and Christiansen [5]; Hartmann and Christiansen [18]; Barndorff-Nielsen [2]; Barndorff-Nielsen and Sørensen [7]).

Here we shall illustrate this further by means of a large data set (Hartmann [16]) collected on the Mediterranean coast of Israel. The data are from different — but related — dynamical environments, and for each environment the variation of the invariant hyperbolic parameters χ (asymmetry), ξ (steepness), and ϱ (tilt) are considered. It is found that among these parameters the most informative is, generally, the tilt ϱ and that the distributions of ϱ values within environments can be well fitted by the beta distribution model. To bring out the information in ϱ most clearly it was found useful to construct a *beta shape triangle*, in analogy with the hyperbolic shape triangle. This construction is presented in section 2. The field and laboratory methods are outlined in section 3. The statistical analysis was performed at two levels: 1) General hyperbolic analysis of the size distributions of the samples and 2) A more detailed study of the distributions within environments of the estimated parameters found on level 1. Levels 1 and 2 are treated in sections 4 and 5, respectively. Section 6 contains some concluding remarks.

2 Shape triangles for the hyperbolic, generalized logistic, and beta distributions

The hyperbolic distribution requires four parameters for its specification and these can be chosen in a variety of ways. Usually two of the parameters are taken to specify the location and the scale (in the mathematical sense) of the distribution. The other two may be chosen to be invariant under location and scale transformations of the underlying variable (such as logarithmic grain size) and the invariant parameters can be said to specify the shape of the hyperbolic distribution. A special choice of two invariant parameters was introduced in Barndorff-Nielsen et al. [4] and is denoted by (χ, ξ). The domain of variation of (χ, ξ) is a triangle (Fig. 1) which is referred to as the hyperbolic shape triangle.

The model function of the hyperbolic distribution, as defined in Barndorff-Nielsen [1], is

$$p(x; \mu, \delta, \phi, \gamma) = a(\delta, \phi, \gamma) \, e^{-1/2(\phi h_- + \gamma h_+)}. \tag{1}$$

Here x indicates the observed variate; μ, δ, ϕ and γ are parameters;

$$h_\pm = \sqrt{\{\delta^2 + (x - \mu)^2\}} \pm (x - \mu); \tag{2}$$

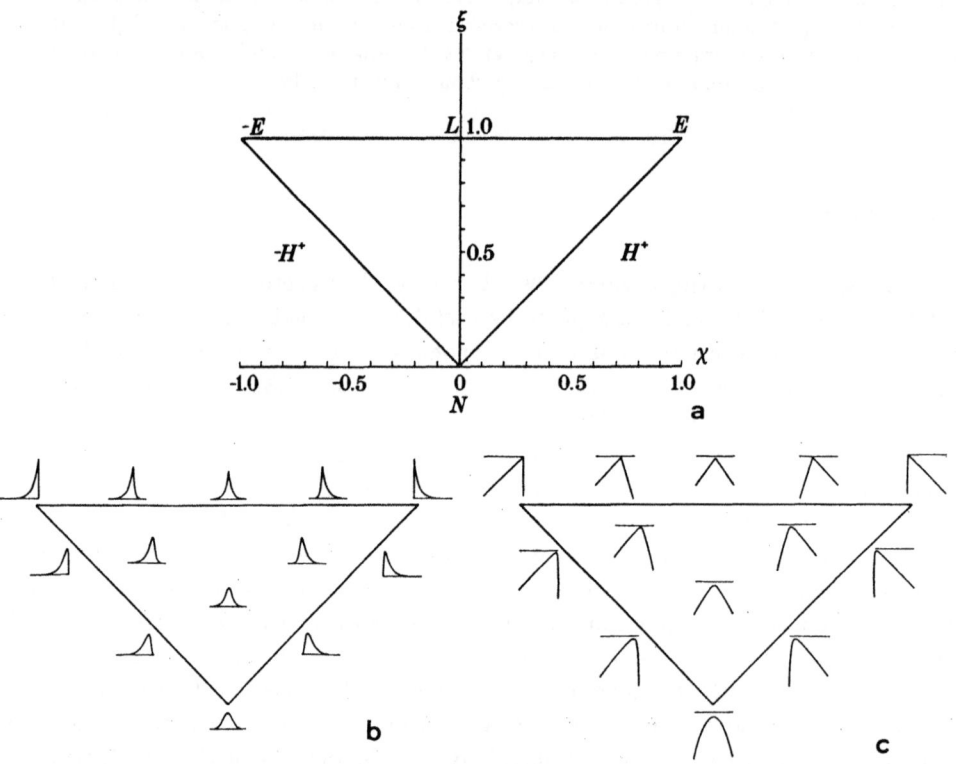

Fig. 1. a The hyperbolic shape triangle, i.e. the domain of variation of the invariant parameters χ and ξ of the hyperbolic distribution. The letters at the boundaries indicate how the normal distribution (N), the positive and negative hyperbolic distributions (H^+ and $-H^+$), the Laplace distribution (symmetric or skew) (L), and the exponential distribution (E) are limits of the hyperbolic distribution. **b** Representative probability functions corresponding to selected (χ, ξ) values, including limiting forms of the hyperbolic distribution. The distributions have been selected so as to have variance equal to unity. **c** The logarithmic probability functions corresponding to (b). (Reproduced from Barndorff-Nielsen and Christiansen [5])

and

$$a(\delta, \phi, \gamma) = \sqrt{(\phi\gamma)}\Big/\Big[\delta(\phi + \gamma)\, K_1\big\{\delta \sqrt{(\phi\gamma)}\big\}\Big], \tag{3}$$

K_1 being a Bessel function. Thus, for fixed values of μ, δ, ϕ and γ, formula (1) determines a probability (density) function on the real line. The invariant parameters χ and ξ are defined by

$$\chi = \{(\phi - \gamma)/(\phi + \gamma)\}\, \xi \tag{4}$$

$$\xi = \big\{1 + \delta \sqrt{(\phi\gamma)}\big\}^{-1/2}, \tag{5}$$

and these are, respectively, termed the *asymmetry* and the *steepness* of the hyperbolic distribution. It is interesting also to consider the parameter ϱ defined by

$$\varrho = \frac{\phi - \gamma}{\phi + \gamma}. \tag{6}$$

We shall refer to ϱ as the *tilt* of the hyperbolic distribution. Note that

$$\chi = \varrho\xi. \tag{7}$$

Any point in the hyperbolic shape triangle determines a particular hyperbolic distribution or one of its limiting forms, as shown in Fig. 1. Furthermore, for not too large values of ϱ one has roughly $\gamma_1 = 3\chi$ and $\gamma_2 = 3\xi^2$, where $\gamma_1 = \varkappa_3/\varkappa_2^{3/2}$ and $\gamma_2 = \varkappa_4/\varkappa_2^2$ aret the standardized skewness and kurtosis of the hyperbolic distribution, cf. Barndorff-Nielsen et al. [4]. Note that although χ and ξ are easily calculated from ϕ, γ and δ, the quantities γ_1 and γ_2 are complicated functions of these three parameters (cf. Barndorff-Nielsen and Blæsild [3]). It is evident from Fig. 1 that throughout the shape triangle, i.e. not just for small to moderate ϱ-values, the parameters χ and ξ express, respectively, the skewness and the kurtosis — in the qualitative sense of these terms — of the hyperbolic distribution.

As will be discussed in sections 4 and 5, certain sedimentological results led to the question of whether it would be possible to define a 'beta shape triangle', i.e. a triangle analogous to the hyperbolic shape triangle but in which the points represent the shapes of the various probability distributions on the interval (0, 1) that are known as beta distributions These distributions have probability density functions of the form

$$\frac{\Gamma(\eta + \theta)}{\Gamma(\eta)\,\Gamma(\theta)}\, x^{\eta-1}(1 - x)^{\theta-1}, \tag{8}$$

$\eta > 0$ and $\theta > 0$ being the parameters.

The beta distributions may be transformed to distributions on the whole real line by the logistic transformation

$$x \to \log \frac{x}{1 - x}. \tag{9}$$

Specifically, the mapping (9) transforms (8) into the probability density function

$$\frac{\Gamma(\eta + \theta)}{\Gamma(\eta)\,\Gamma(\theta)}\, e^{\eta x}(1 + e^x)^{-(\eta+\theta)} \tag{10}$$

where $-\infty < x < \infty$. Like the hyperbolic distribution, (10) has log linear tails, and this remains true if we further make a location-scale transformation

$$x \rightarrow \delta x + \mu,$$

which turns (10) into

$$\frac{\Gamma(\eta + \theta)}{\Gamma(\eta)\,\Gamma(\theta)}\,\delta^{-1}e^{\eta(x-\mu)/\delta}\,\{1 + e^{(x-\mu)/\delta}\}^{-(\eta+\theta)}. \qquad (11)$$

In fact, any hyperbolic distribution can be very closely approximated by one of the distributions of the form (11). The latter are referred to as generalized logistic distributions and they have been shown (Barndorff-Nielsen et al. [6]) to have a number of probabilistic properties very similar to those of the hyperbolic distributions.

In view of the close analogy between the hyperbolic distributions and the generalized logistic distributions, we may now define a shape triangle for the latter by directly mimicking the definition of the hyperbolic shape triangle. Due to the abovementioned relation between the beta distributions and the generalized logistic distributions this 'logistic shape triangle' may also be viewed as a 'beta shape triangle'.

Specifically, we define the logistic shape triangle as the domain of joint variation of the parameters

$$\xi' = \left\{1 + \sqrt{(\eta\theta)}\right\}^{-1/2} \qquad (12)$$

and

$$\chi' = \varrho'\xi' \qquad (13)$$

where

$$\varrho' = \frac{\eta - \theta}{\eta + \theta}. \qquad (14)$$

These definitions of ξ', χ' and ϱ' are indeed quite analogous to those for the hyperbolic distributions because the slopes of the log linear tails of (4) are $\varphi = \theta/\delta$ and $\gamma = \eta/\delta$, respectively.

The beta shape triangle (Fig. 2) is then defined as the domain of joint variation of ξ' and χ' where these variables are also given by (12) and (13), with η and θ from (8) and with $\delta = 1$. We also use ϱ', given by (14), in the context of the beta distribution.

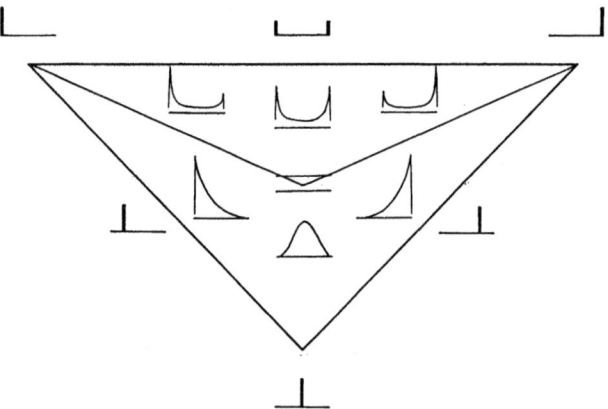

Fig. 2. The beta shape triangle showing representative probability functions corresponding to selected (χ, ξ) values, including limiting forms of the beta distribution

3 Field and laboratory methods

In this study we adopt a population concept in line with Miller [22], Greenwood [15] and Hartmann [16]. In principle it is possible to define a spot sample of sand as a population. Here, however, the target population is all the grains in the active layer of the littoral environments in the study area. This population is not in practice available. Therefore many samples were collected in the different environments in order to assess the local sedimentological variation associated with single samples.

A total of 937 sediment samples were collected in transects from the active layer (Otto [23]) along a 54 km stretch of the nontidal Southern Mediterranean coast of Israel. Sampling positions were 0) Deep inshore 1) Shallow inshore 2) Step 3) Mid swash zone 4) Top swash zone 5) Backshore 6) Far backshore, and D) Coastal dunes. See Hartmann ([17], Fig. 1 and 2) for location of study area and general sketches of profiles. The transects were separated by intervals of 200 m. In addition, 175 samples were collected from the beach and the inshore at the Herzlia beach.

After removal of salt with freshwater, 3 to 3.5 g splits of the samples were analysed by an automated settling tube system (Goldbery and Tehori [14]). Settling velocity was used directly as a measure of size in accordance with Reed et al. [24] and Bryant [10].

Hyperbolic grain-size parameters were estimated by the method of maximum likelihood, using the SAHARA program (Christiansen and Hartmann [11]). Further, for each of the dynamical environments, the distribution of the estimates of the tilt parameter ϱ from the individual hyperbolic distributions were fitted by the beta distribution model, also by maximum likelihood.

4 Size distributions

For each of the dynamical environments the estimated (χ, ξ) values were plotted in the hyperbolic shape triangle and contoured according to a method presented in Christiansen and Hartmann [12]. There are considerable overlaps between the distributions of the estimated (χ, ξ) values from the different environments. However, each of the distributions differs significantly from the others. In Fig. 3 we have plotted, for each of the environments, the region which has the highest density of χ, ξ values.

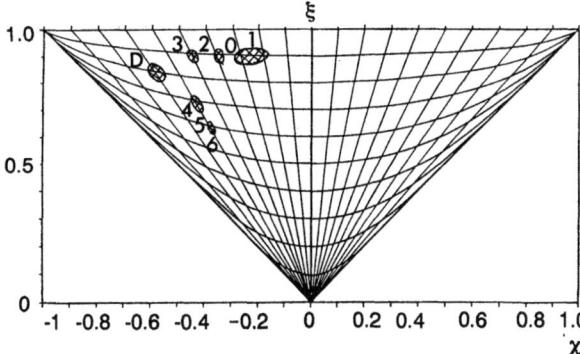

Fig. 3. The hyperbolic shape triangle. Straight lines indicate $\varkappa$-erosion/deposition. Hammock-shaped curves correspond to ε-erosion/deposition. Areas of highest density of (χ, ξ) values are also shown. Numbers refer to the littoral environments, as in Hartmann ([17], Fig. 2)

Also shown in that figure are the curves corresponding to pure ε-erosion/deposition, i.e. the hammock-like curves, and the curves, in fact straight lines, corresponding to pure $\varkappa$-erosion/deposition. (The concepts of pure ε- and $\varkappa$-erosion/deposition are elements of a model for erosion and deposition developed in Barndorff-Nielsen and Christiansen [5] and Barndorff-Nielsen and Sørensen [7], and we refer the reader to these papers for details.) These two sets of curves may be viewed as determining an intrinsic coordinate system in the hyperbolic shape triangle. One of the coordinates in this system is the tilt ϱ and the other, which we shall denote by σ, may be taken to be the ξ-coordinate of the point where the hammock curve in question intersects the ordinate axis. The formula for the second coordinate, expressed in terms of the original hyperbolic parameters, is

$$\sigma = \{1 + \delta(\varphi + \gamma)/2\}^{-1/2}. \tag{15}$$

A visual inspection of Fig. 3 shows that the subaqueous environments follow one trend and that the subaerial environments follow another, Hartmann [17]. The two trends indicate that ε-erosion/deposition dominates in the subaqueous environment and that the subaerial environments primarily follow $\varkappa$-erosion/deposition. Such trends can also be found in Barndorff-Nielsen and Christiansen ([5], Fig. 5 and 7). ε-erosion/deposition depends mainly on the available energy while $\varkappa$-erosion/deposition takes into account the influence of the differences in size between the sand particles in the bed. A possible explanation of the difference in trends could therefore be the differences in flow dynamics and density between wind and water, in consequence of which the saltation collision process is of much greater import in the subaerial environments than in the subaqueous environments.

Note, in particular, that not only the tilt parameter ϱ but also the steepness parameter ξ carries substantial information about the dynamical character of the environment, as illustrated by Fig. 3. (In investigations by Blatt et al. [8] and McLaren [21] kurtosis values were found to be inconclusive for environmental discrimination. The difference is, we propose, due primarily to the fact that parametric modelling by the hyperbolic distribution is much more incisive than the use of empirical moments.)

5 Distributions of invariant parameters

The (estimated) parameters turned out to have a substantial range of distributional forms depending on the environment, as indicated already in the previous section. Furthermore, it was found that the beta distribution fitted very well to most of the marginal distributions of the invariant parameters. See Fig. 4 which illustrates this for the tilt parameter ϱ.

Figure 5 exhibits, for each of the eight environments, the relation between the estimated values of the invariant parameters ϱ, χ and ξ. Note that, in view of the equation (7), we necessarily have

$$\varrho < \chi < 0 \quad \text{if} \quad \varrho < 0$$

$$0 < \chi < \varrho \quad \text{if} \quad \varrho > 0,$$

as also indicated by the diagonal lines in Fig. 5.

The main conclusion from Fig. 5 is that within each environment there is no populational correlation between ϱ and σ. (We are relying here on the fact that plots of ϱ against σ would have been quite similar to the plots, in Fig. 5, of ϱ against ξ, cf. also Fig. 3.) This means

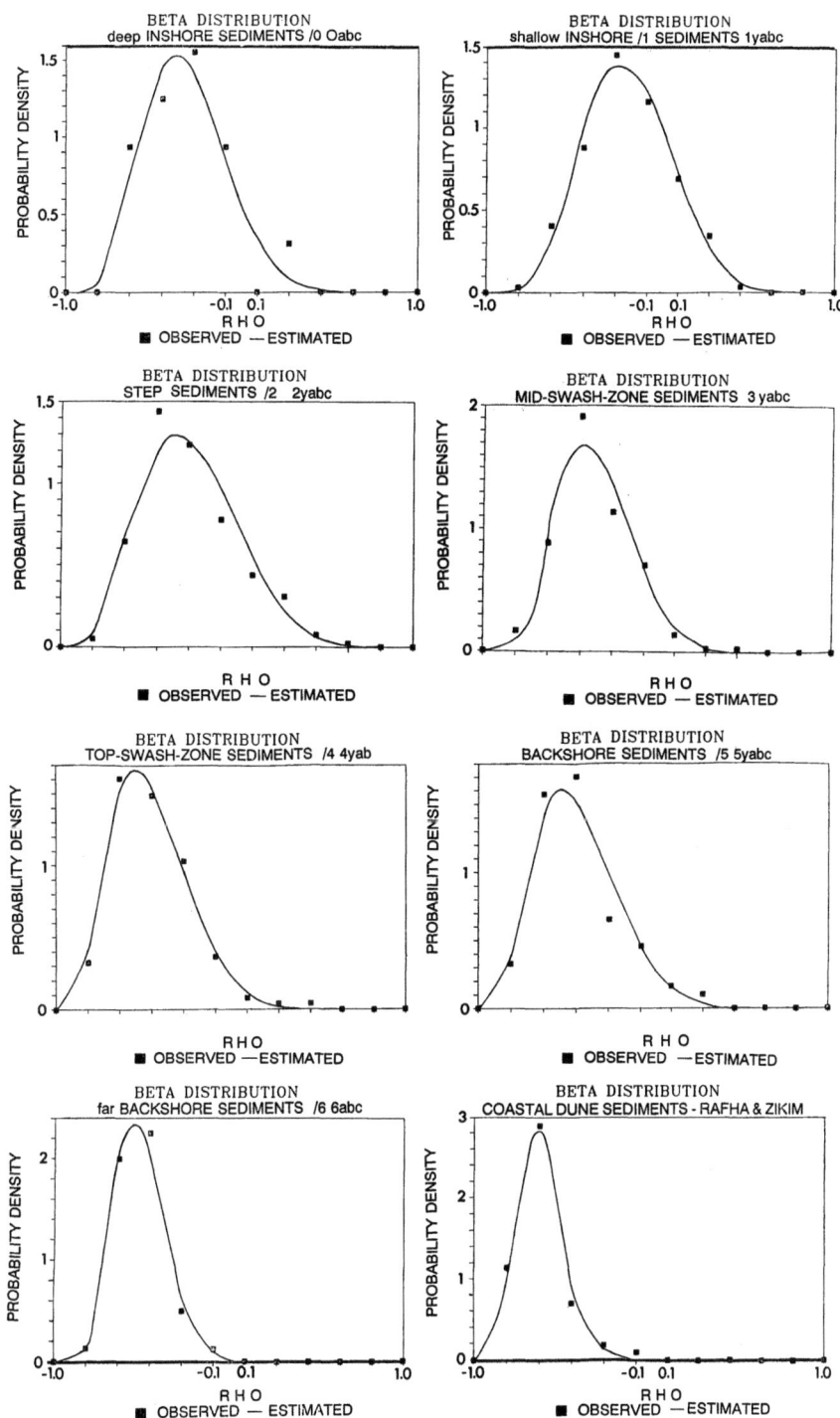

Fig. 4. Distributions of ϱ in the different subenvironments together with the fitted beta distributions

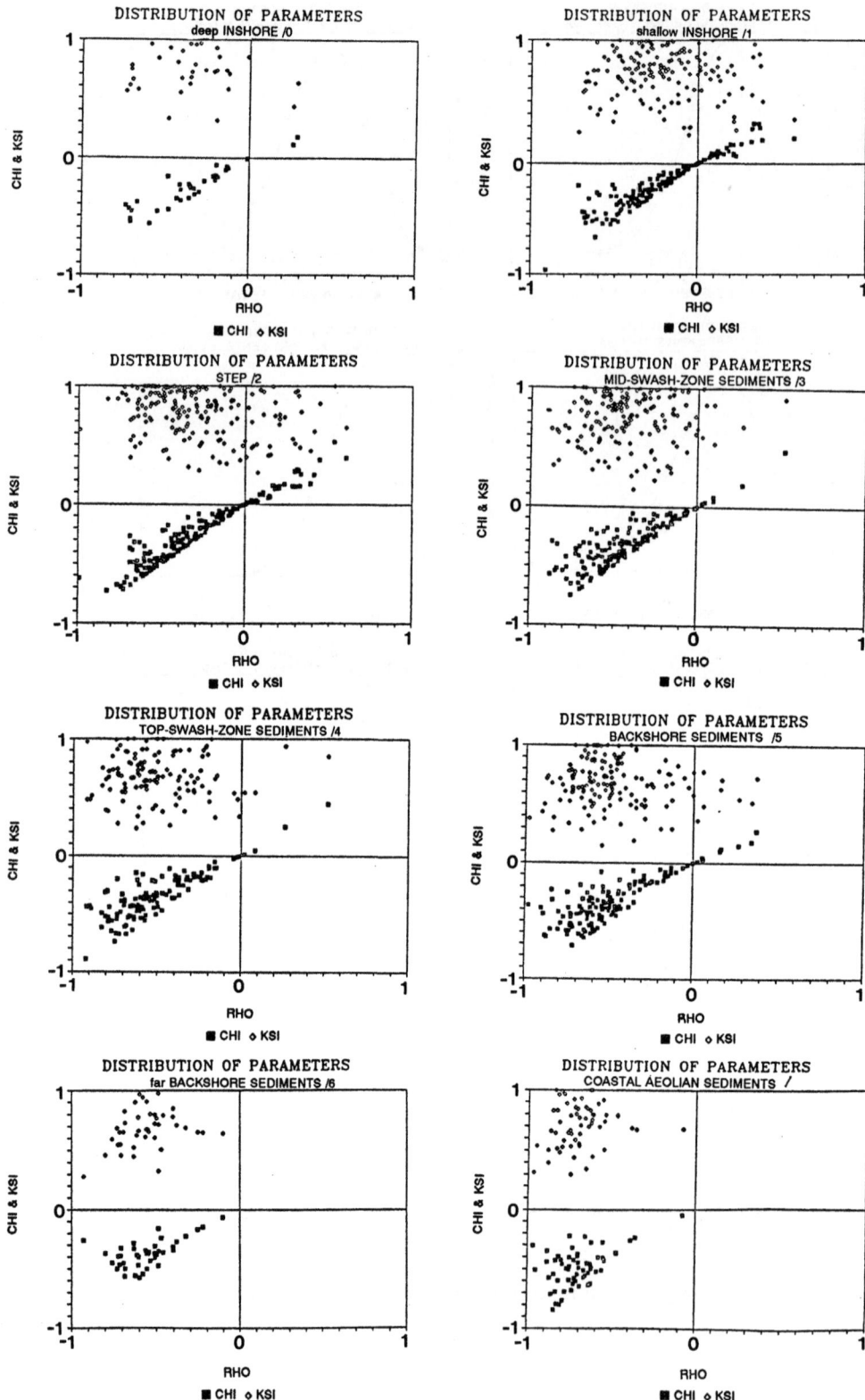

Fig. 5. Bivariate plots showing relations between ϱ, χ and ξ

that, essentially, the statistical analysis of the variation of the hyperbolic shape parameters breaks up into two independent parts, one for ϱ and one for σ. It may be that this observed independence of ϱ and σ is a reflection of (approximate) orthogonality of ϱ and σ in the sense of asymptotic likelihood theory, a possibility that calls for further investigation.

Figure 6 shows the positions in the beta shape triangle of the fitted distributions of hyperbolic tilt from the eight different environments (cf. Fig. 4). The study of the hyperbolic shape triangle showed that the hyperbolic tilt ϱ is that among the shape parameters which carries most information about the shape of single distributions. The same phenomenon but now for the beta tilt ϱ' is emerging in the beta shape triangles for the populations. In Fig. 7 the nine environments are sorted according to ascending ϱ' values starting with the most symmetrical distributions of the shallow inshore and ending with the most asymmetrical distribution of the coastal dune environment. It appears from Fig. 7 that, again, we can distinguish two main groups, the subaqueous and the subaerial environments.

The 197 samples from the beach and the inshore at Herzlia do not fit into this pattern. Hartmann [16] showed that sediments at Herzlia were the end-product of the longshore transport in the Nile littoral cell. Therefore the sediments are very well sorted, homogeneous, fine to very fine sands without any difference in grain size distributions between the different littoral environments. Hartmann [16] concluded that the sediments at Herzlia were of such a narrow range that no further sorting differentiation could take place.

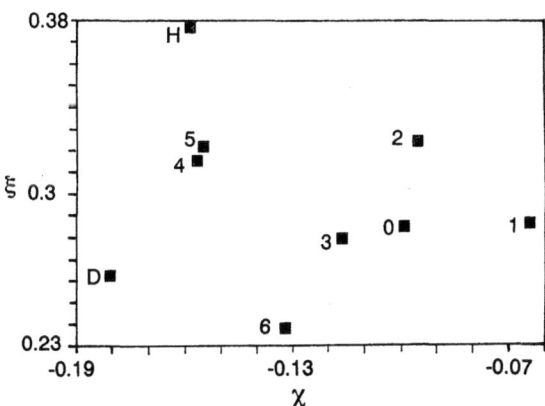

Fig. 6. Section of the beta shape triangle with the estimated (χ, ξ)-values from the different littoral environments

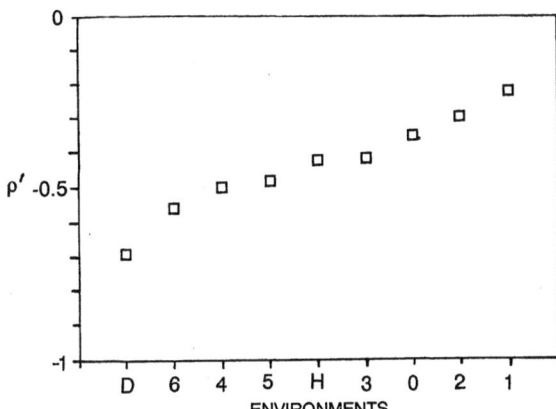

Fig. 7. Values of ϱ' for the different environments, sorted according to ascending ϱ'-values. Abscissal tags refer to the littoral environments

6 Discussion

From a profile perpendicular to the shore line Fox et al. [13] found a marked difference in both kurtosis and skewness values between the subaqueous and the subaerial parts of the profile. From four traverses across beach, dune and aeolian flats Mason and Folk [20] concluded that the differences between the environments were almost entirely in the tails of the distributions which affected skewness and kurtosis. The same conclusions can be drawn from Fig. 7 which are based on 270 traverses across the littoral subenvironments, see Hartmann [16, 17]. The present paper for the first time shows that the same phenomenon can be observed in the distributions of the shape parameters ϱ and ϱ'. Such observations might help distinguishing former coastlines in deposits where internal structure and traditional grain size analysis are inconclusive in environmental discrimination (Hobday and Horne [19]; Bowman et al. [9]).

References

[1] Barndorff-Nielsen, O. E.: Exponentially decreasing distributions for the logarithm of particle size. Proc. R. Soc. Lond. A **353**, 401—419 (1977).

[2] Barndorff-Nielsen, O. E.: Sorting, texture and structure. Proc. R. Soc. Edinburgh **96 B**, 167—179 (1989).

[3] Barndorff-Nielsen, O. E., Blæsild, P.: Hyperbolic distributions and ramifications: contributions to theory and applications. In: Statistical distributions in scientific work, vol. 4. (Taillie, C., Patil, G. P., Baldessari B. A.,) pp. 19—44, Dordrecht: D. Reidel, (1981).

[4] Barndorff-Nielsen, O. E., Blæsild, P., Jensen. J. L., Sørensen, M.: The fascination of sand. In: A celebration of statistics. (Atkinson, A. C., Fienberg, S. E., eds.) pp. 57—87 New York: Springer (1985).

[5] Barndorff-Nielsen, O. E., Christiansen, C.: Erosion, deposition and size distributions of sand. Proc. R. Soc. London A **417**, 335—352 (1988).

[6] Barndorff-Nielsen, O. E., Kent, J., Sørensen, M.: Normal variance-mean mixtures and z distributions. Internat. Statist. Rev. **50**, 145—159 (1982).

[7] Barndorff-Nielsen, O. E., Sørensen, M.: On the temporal-spatial variation of sediment size distributions. Acta Mechanica [Suppl. 2], 23—35 (1991).

[8] Blatt, H., Middleton, G., Murray, R.: Origin of sedimentary rocks. 634 pp. Englewood Cliffs, N. J.: Prentice-Hall (1972).

[9] Bowman, D., Christiansen, C., Magaritz, M.: Late-Holocene coastal evolution in the Hanstholm-Hjardemaal region, NW Denmark. Morphology, sediments and dating. Geogra. Tidss. **89**, 49—57 (1989).

[10] Bryant, E. A.: Sample size variation in settling velocity distribution subpopulations using curve dissection analysis. Sedimentology **33**, 767—775 (1986).

[11] Christiansen, C., Hartmann, D.: SAHARA: a package of PC computer programs for estimating both log-hyperbolic grain size parameters and standard moments. Comp. Geosci. **14**, 557—625 (1988).

[12] Christiansen, C., Hartmann, D.: The hyperbolic distribution. In: Principles, methods and application of particle size analysis. (Syvitski, J. P. M., ed.) Cambridge University Press. (In press) (1990).

[13] Fox, W. T., Ladd, J. W., Martin, M. K.: A profile of the four moment measures perpendicular to a shore line, South Haven, Michigan. J. Sed. Petrol. **36**, 1126—1130 (1966).

[14] Goldbery, R., Tehori, O.: SEDPAK — a comprehensive operational system and data-processing package in APPLESOFT BASIC for a settling tube, sediment analyzer. Comp. Geosci. **13**, 565—585 (1987).

[15] Greenwood, B.: Sediment parameters and environment discrimination: an application of multivariate statistics. Can. J. Earth Sci. **6**, 1347—1358 (1969).

[16] Hartmann, D.: Coastal sands of the sourthern and central part of the Mediterranean coast of Israel — reflection of dynamic sorting processes. Ph. D. Thesis, Faculty of Science, Aarhus University. 335 pp. (1988).

[17] Hartmann, D.: Cross-shore selective sorting processes and grain size distributional shape. Acta Mechanica [Suppl. 2]; 49—63 (1991).

[18] Hartmann, D., Christiansen, C.: Settling velocity distributions and sorting processes on a longitudinal dune: a case study. Earth Surf. Proc. Landf. **13**, 649—656 (1988).

[19] Hobday, D. K., Horne, A. R.: The Port Dunford Formation: a major Pleistocene barrier lagoon complex along the Zululand coast. Trans. Geol. Soc. South Africa **77**, 141—149 (1974).

[20] Mason, C. C., Folk, R. L.: Differentiation of beach, dune and eolian flat environments by size analysis, Mustang Island, Texas. J. Sed. Petrol. **28**, 211—226 (1958).

[21] McLaren, P.: An interpretation of trends in grain size measures. J. Sed. Petrol. **51**, 611—624 (1981).

[22] Miller, R. L.: A model for the analysis of environments of sedimentation. J. Geol. **62**, 108—113 (1954).

[23] Otto, C. H.: The sedimentation unit and its use in field sampling. J. Geol. **46**, 569—582 (1938).

[24] Reed, W. E., LeFever, R., Moir, G. J.: Depositional environment interpretation from settling velocity (Psi) distributions. Geol. Soc. Am. Bull. **86**, 1321—1328 (1975).

Authors' addresses: O. E. Barndorff-Nielsen, Prof., Sc. D., Department of Theoretical Statistics, Institute of Mathematics, and C. Christiansen, Docent, Dr. Sc., Department of Earth Sciences, Aarhus University, Ny Munkegade, DK-8000 Aarhus C, Denmark; D. Hartmann, Ph. D., Israel Oceanographic and Limnological Research Ltd., Tel Shikmona, P.O.B. 8030, Haifa 31 080, Israel.

Acta Mechanica (1991) [Suppl] 2: 49—63
© by Springer-Verlag 1991

Cross-shore selective sorting processes and grain size distributional shape

D. Hartmann, Haifa, Israel

Summary. Applying the population concept on more than a thousand beach sand samples from the southern Israeli Mediterranean coast allowed abandonment of the common single sample approach. The study suggests six distinct grain size populations related to a sequence of six beach environments, entitled: 0. and 1. inshore; 2. step; 3. mid-swash zone; 4. top-swash zone; 5. backshore; 6. and D. far-backshore wind blown sand and coastal dunes. The invariant hyperbolic shape parameter ϱ indicates a gradual change of the grain-size distributional form across the left part of the hyperbolic shape triangle. The shape positions (χ, ξ) in the triangle suggest that the subaqueous populations are subjected mainly to — erosion-deposition processes and move along one of the upper 'hammock' curves in the shape triangle. The subaerial populations were found to follow mainly $\varkappa$-erosion-deposition processes and to move along the -0.6 to -0.7 ϱ lines. However, this group is divided into two basically different depositional environments: water-lain backshore populations dominated by the swash-backwash bidirectional sheet flow and wind-blown sand originating from the backshore sediments. The different grain-size populations reflect the sum of modes of transport which dictate the grain size cutoffs and the typical grain size, thus defining the shape of the distributions.

1 Introduction

Studies of sediment sorting processes normal to the beach have been performed on different beaches all over the world with varying results (Evans [1], Miller and Zeigler [2], Seibold [3], Schiffman [4], Fox et al. [5], Ingle [6], Greenwood and Davidson-Arnott [7], Bryant [8]). The spatial and temporal sediment-size patterns were related to changes in energy levels and flow directions. Nevertheless, Komar [9] (p. 350) concludes that "..., the processes responsible for the selective sorting of grain sizes across the profile remain the most poorly documented and understood facets of nearshore processes." Many of the sorting processes, especially those on beaches with a very wide range of grain-sizes (Bluck [10]), are obvious and the observations could be accomplished in the field without using any sophisticated analytical techniques. In many other cases, it seems that the sand does not reflect the prevailing hydrodynamic conditions. Most of the studies were performed by the analysis of samples, and the approaches were relatively crude.

A common technique in sedimentology is to compare samples by performing statistical analyses. This method can succeed if we work in environments in which we know *a priori* that the samples are different and there is a relatively wide range of grain-sizes available for the different dynamic sorting processes. As most of the investigations used sieving techniques, the number of samples analyzed was kept low, usually at most a few dozens. The more complicated cases usually turned out to be indeterminate (Pranzini [11]). In any effort to generalize the findings from one place to another, exceptions and even reverse behaviour were often encountered (Sedimentation Seminar [12]).

The development of low-cost, automatically recording settling tube systems for the hydraulic grain size analysis of sand (Gibbs [13], Geldof and Slot [14]) allows us to increase markedly the number of samples analyzed and the quality of the grain-'size' property (Hartmann [15]). However, a larger number of samples introduces more sedimentological noise and as a result one sometimes does not see the forest because of too many trees (Kachholtz [16]).

Most of the sedimentological systems can be considered as process response models (Krumbein and Graybill [17]). The coastal environments and their components, i.e., forces, grains and morphology, have — like other natural phenomena — a probabilistic nature. These systems are subjected to varying forces and show a variety of forms and compositions. For the study of such systems we suggest abandonment of the single sample approach and use a concept already applied by Russel [18] and Shea [19]. This 'population' concept considers single samples as random components of a larger population and looks at the distribution of the composite samples as bearing the sedimentological information.

One of the results of the population concept is the recognition that the estimated parameters for the samples which build the populations, and quantities derived from the estimates, also follow a probabilistic nature and that the beta distribution model can be used to describe them (Hartmann [15], Barndorff-Nielsen et al. [20]).

2 Study area

The study area consists of a 60 km segment along the Mediterranean beach from the Egyptian-Israeli border at Rafah to the Barnea Beach at Ashqelon (Fig. 1). The southern Israeli Mediterranean coast is a part of the Nile littoral cell, and any understanding of the coastal processes in that coast region should be regarded in this frame. It is a straight, plane shaped and medium to fine-sized sandy beach, backed by coastal dunes with sporadic low coastal eolianite cliffs. The study area is exposed to a microtidal range averaging less than 50 cm.

The prevailing wave characteristics during the autumn study period were $H_{sig} = 0.3$ to 1.3 m, $H_{max} = 0.4 - 1.9$ m and $T_{sig} = 3.0 - 7.0$ s, with a W−WNW incident wave direction. The prevailing wave characteristics during the spring were $H_{sig} = 0.1 - 2.0$ m, $H_{max} = 0.1 - 3.0$ m and $T_{sig} = 4.5 - 10.0$ s, with a W−WNW incident wave direction. The autumn (segment A) and the spring (segment D2) beach profiles (Fig. 2) differ significantly as well as does the pattern of the nearshore circulation. During the autumn the inshore shows a well-developed rhythmic pattern of rip-current cells. The spring survey revealed a distinct longshore current to the north.

3 Field and laboratory methods

The surveyed beach segments (Fig. 1) were from Rafah to Ashqelon (segment A) and from Erez to Ashqelon (segment D2). Sampling was carried along shore-normal transects. Distance between the transects was 200 meters. 285 transects were sampled on a 'storm' beach (A), and 78 were sampled on an 'accumulating' beach (D2). Sediments were taken in a systematic stratified sampling survey from eight defined beach environments: 0) = deep inshore, 1) = shallow inshore, 2) = step, 3) = mid-swash zone, 4) = top-swash zone,

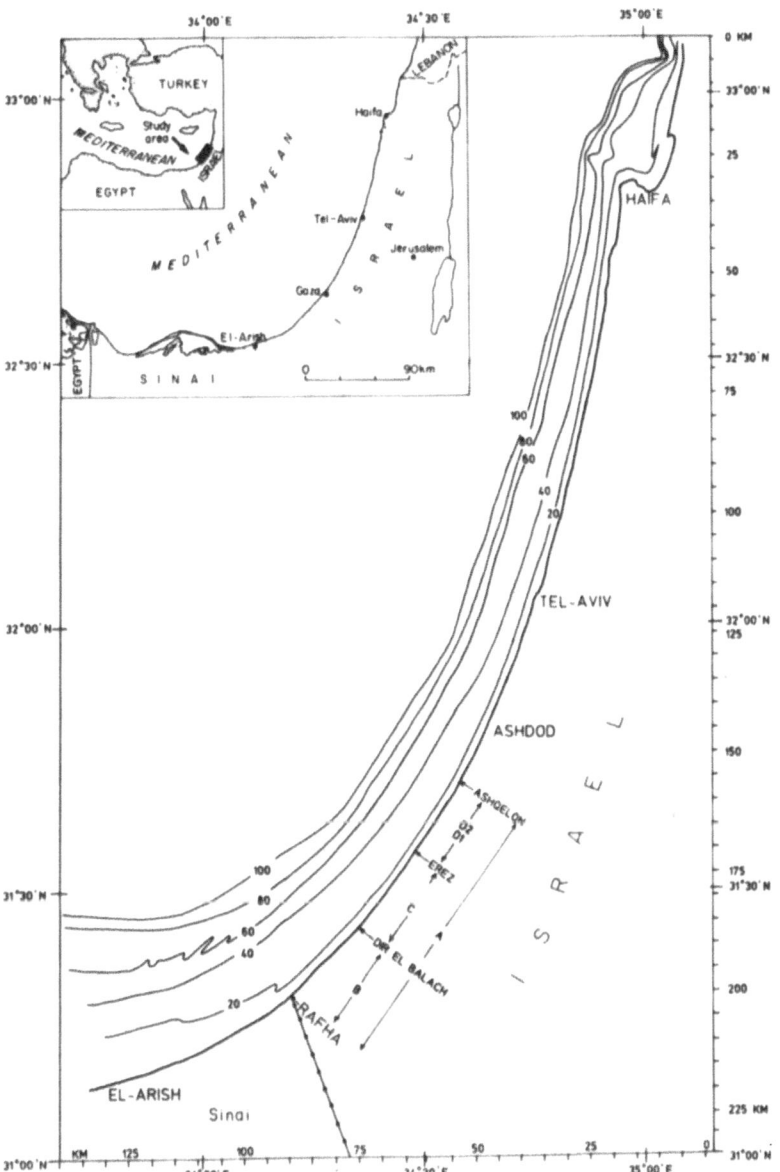

Fig. 1. Location maps showing study area on the Mediterranean coast of Israel and the beach segments surveyed

5) = backshore, 6) = beach wind blown sand and D) = coastal dunes. Two inshore environments (deep inshore and shallow inshore) are defined more generally because of rapidly changing bottom topography and accessibility for sampling. Average beach profiles along each transect were constructed and studied (Fig. 2). This means that the investigation has a rather low and general level of resolution. As the different environments are well-defined and categorized, the paper attempts to describe the sediments in a generalized beach profile localized in time and space. These profiles are used to explain the general cross-shore selective sorting processes and to demonstrate that they act as sorting differentiation processes.

About two thousand samples were analyzed by an automated settling tube built after

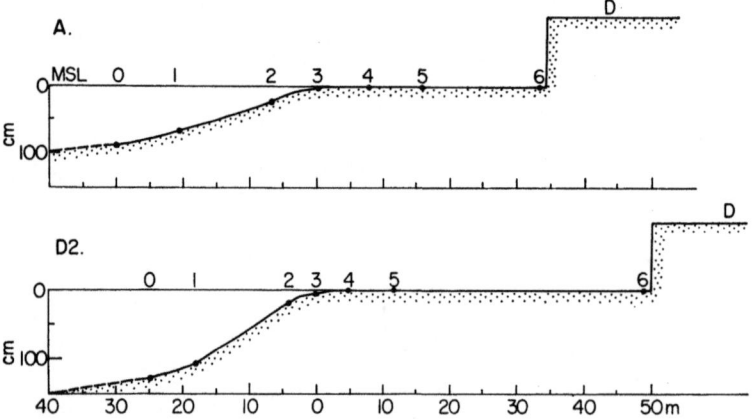

Fig. 2. Generalized cross-shore profiles with location of the sampling positions indicated. **a)** segment A September 1983, **b)** segment D2 April 1984

Mayo [21] and documented by Goldbery and Tehory [22]. In this study the finite settling velocity serves as the grain-'size' attribute in accordance with Reed et al. [23], May [24] and Bryant [25]. Through the present study, the word 'grain-size' will be used as equivalent to 'settling velocity'.

Empirical grain size probability distributions were constructed by summing up all the single samples of a given localized environment and are referred to as super samples. They can be generated by physical or computational blending and are regarded as representing natural populations of sediments. Their probabilistic nature is assumed to throw light on the cross-shore selective sorting patterns in the study area.

Hyperbolic grain size parameters, the four standard moments and some other textural parameters were estimated and calculated with the SAHARA program (Christiansen and Hartmann [26] and Jensen [27]). The distribution of the shape positions (χ, ξ) in the hyperbolic shape triangle was studied graphically and statistically by using newly developed tools to present, evaluate and compare hyperbolic shape position data for analyzed grain-size populations (Hartmann [15] and Hartmann and Christiansen [28]). The graphical method of contouring the scatter diagrams of the samples in the domain of the hyperbolic shape triangle and chi-squared tests were performed on the sequence of 8 environments normal to the beach. The investigated samples are from the entire Rafah—Ashqelon beach segment and represent the autumn profile.

4 The hyperbolic shape triangle

Following Barndorff-Nielsen et al. [29], the domain of variation of the hyperbolic invariant parameters χ (asymmetry) and ξ (steepness) is a triangle which is referred to as the hyperbolic shape triangle (Barndorff-Nielsen et al. [20], Fig. 1). Barndorff-Nielsen and Blæsild [30] demonstrated that throughout the hyperbolic shape triangle, χ and ξ express, in the qualitative sense of these terms, the skewness and kurtosis of the hyperbolic distribution, respectively. Any pair of these invariant parameters determines a particular shape of a hyperbolic distribution or one of its limiting forms.

The shape elements of the distributions are usefully expressed by the combined shape position (χ, ξ) localized in the hyperbolic shape triangle. The parameter ξ is very sensitive to various types of noise and often exhibits in empirical populations a highly stochastic tendency (Hartmann [15] and Barndorff-Nielsen et al. [20]). The ratio $\varrho = \chi/\xi$, which is termed the tilt parameter, has turned out generally to contain the best information about the shape of a distribution.

Hartmann [15] and Barndorff-Nielsen et al. [20], Fig. 5 this volume show the relationship between ϱ and the two invariant shape parameters χ and ξ. Even a visual inspection of the figures shows that there is no correlation between the parameters ξ and ϱ (whereas there is always a significant correlation between χ and ϱ which is based on the functional inequalities $\varrho \gtreqless \chi$ according as $\varrho \gtreqless O$).

Bagnold and Barndorff-Nielsen [31] briefly presented a dynamical model for the generation of hyperbolic curves. Barndorff-Nielsen and Christiansen [32] presented a mathematical-physical model for the net effect of erosion and deposition of sedimentary populations of sand grains. According to them the 'hammock' curves (see Fig. 6 below) are related to pure ε erosion-deposition and the straight lines to pure $\varkappa$ erosion-deposition. For a given hyperbolic distribution, deposition will lead towards the left hand part of the shape triangle and erosion towards the right hand part. The fact that entrainment and deposition changes the distributions over a conitnuum of shape forms makes the model an excellent sedimentological working tool. This paper will concentrate solely on the shape parameters of the grain size distributions.

The pattern of the estimated shape positions (χ, ξ) in the hyperbolic shape triangle can be quantified and mapped (Hartmann [15] and Hartmann and Christiansen [28]). Two procedures were used for investigating the behaviour of the shape positions of individual samples from different super-samples. One technique is quantitative — a chi-squared test comparing a pair of environments at a time. The second technique is point density mapping and qualitative pattern recognition. The distributions of the shape positions in the nonlinear $(\varepsilon, \varkappa)$ space of the hyperbolic shape triangle determine qualitatively the selective sorting processes normal to the shore of the sediments in the study area.

5 Results

5.1 Field data and beach morphology

The generalized beach profile representing the beach segment Rafah to Ashqelon (A) was taken at the end of September 1983. It is a storm profile or profile indicating an early accretional stage. In contrast, the generalized beach profile Erez to Ashqelon (D2), sampled at the beginning of April 1984, is steep and represents accumulation with maximum sediment storage on the beach causing high reflectivity (Fig. 2). However, the profile data show that the general autumn profile consists of a subaqueous storm profile superimposed on a well-developed subaerial summer profile, with most of the sediment still stored in the backshore. The spring profile represents accumulation on the berm, but the subaerial profile still bears the characteristics of storms.

5.2 Hyperbolic shape parameters

All the grain-size distributions were estimated according to the log-hyperbolic and the log-normal distributional models. The hyperbolic model permits representation of the wide diversity of distributional shapes over the entire size interval while the log-normal

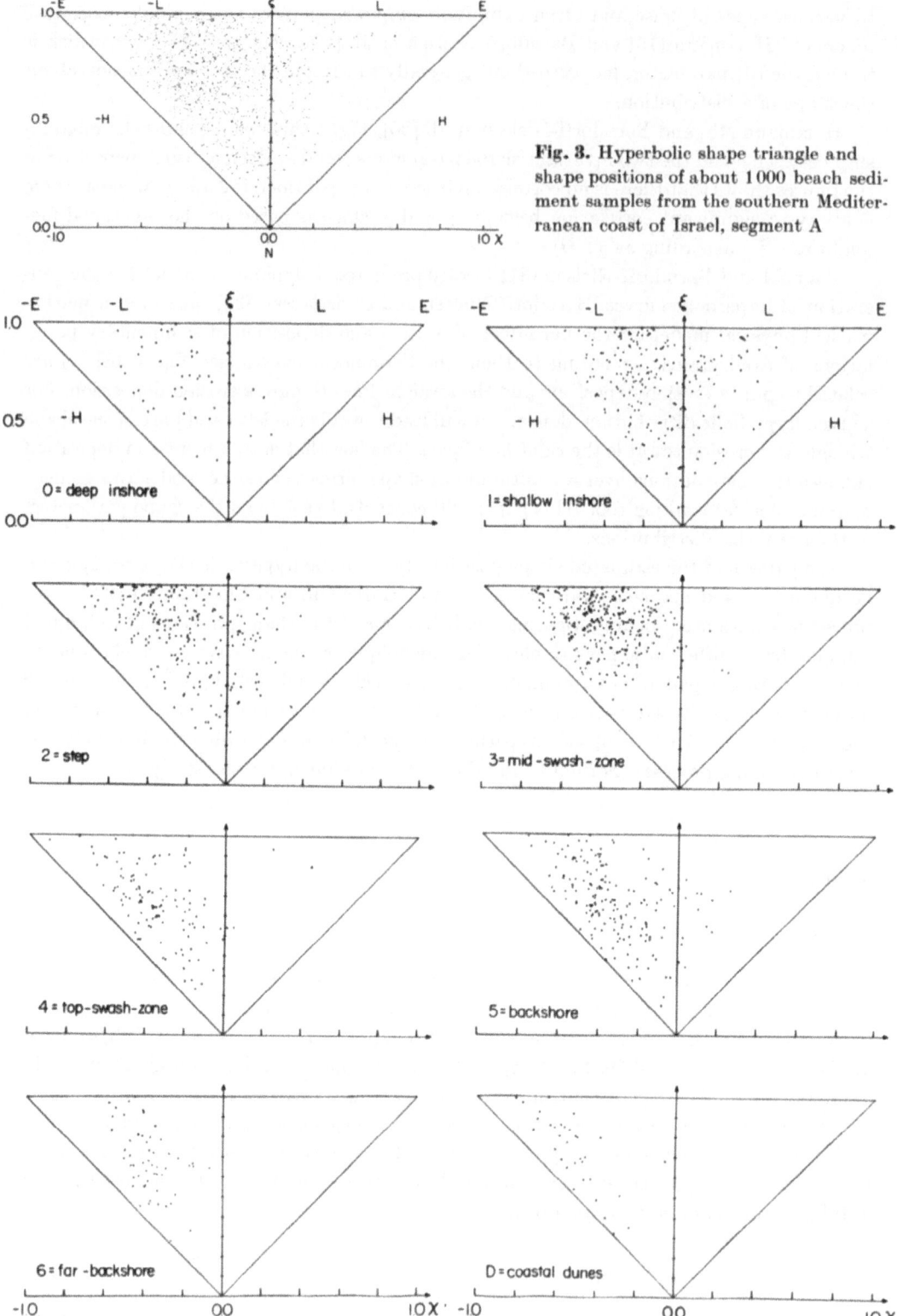

Fig. 3. Hyperbolic shape triangle and shape positions of about 1000 beach sediment samples from the southern Mediterranean coast of Israel, segment A

Fig. 4. Shape positions of the eight cross-shore environments from segment A shown in hyperbolic shape triangles

model could be rejected for most of the samples (Fig. 3). Figure 4 presents the shape positions of the sequence of eight cross-shore environments. The estimated hyperbolic shape parameters χ, ξ, and ϱ constitute the studied database. Vincent [43] showed that $\pi = \varrho / \sqrt{(1 - \varrho^2)}$ is an important parameter for environmental discrimination. As the tilt parameter ϱ can be followed up in the hyperbolic shape triangle, it is more useful as a working tool.

5.2.1 Shape positions (χ, ξ) distributions - chi-square test

This method has been used on the information presented in Fig. 5. Table 1 (see Hartmann [15] for more details) shows the results of the chi-square tests which were performed on all the possible combinations of pairs of beach environments in the Rafah-Ashqelon beach segment. It is possible to distinguish 4 different relationship patterns between the environments:

1) $--$: The single samples of the two super-samples have very similar distributions of shape positions. The test criterion fails totally to achieve significance.
2) $-$: Similar distributions of shape position. The test criterion fails to achieve significance at some specified probability level (0.95).
3) $+$: The single samples of the two super-samples have different distributions of shape positions. It is significant at some specified probability level (0.975).
4) $++$: Totally different distributions of shape positions with significance at a very high probability level (> 0.995).

The results suggest that neighbouring subaerial beach environments show no significant differences. The coastal dune and the far-backshore wind blown sediments have very similar shape position distributions. As the far-backshore samples were taken from the rippled wind blown sand at the end of the backshore, the results are in accordance with field observations. They demonstrate that the shape of the grain-size distributions reflects the physical processes, mainly sorting by wind. The far-backshore samples resemble very much the backshore samples, which shows the mutual connection between them, as the backshore sediments are the source material for the wind blown sand. However, the fact

Table 1. Chi-squared comparisons between the eight beach environments using the occurrence of the shape positions (χ, ξ) in the shape triangle

		0	1	2	3	4	5	6	D
Deep inshore	(0)	\	$--$	$--$	$-$	$+$	$(+)$	$+$	$++$
Shallow inshore	(1)		\	$+$	$++$	$++$	$++$	$++$	$++$
Step	(2)			\	$(+)$	$++$	$++$	$++$	$++$
Mid-swash-zone	(3)				\	$++$	$++$	$++$	$++$
Top-swash-zone	(4)					\	$--$	$+$	$+$
Backshore	(5)						\	$-$	$+$
Top-backshore	(6)							\	$-$
Coastal dunes	(D)								\

$++$ = highly significant difference (> 0.995)
$+$ = significant difference at the level of 0.975
$(+)$ = marginal significant difference at the level of 0.950
$-$ = no significant difference at the level of 0.950
$--$ = high degree of overlap

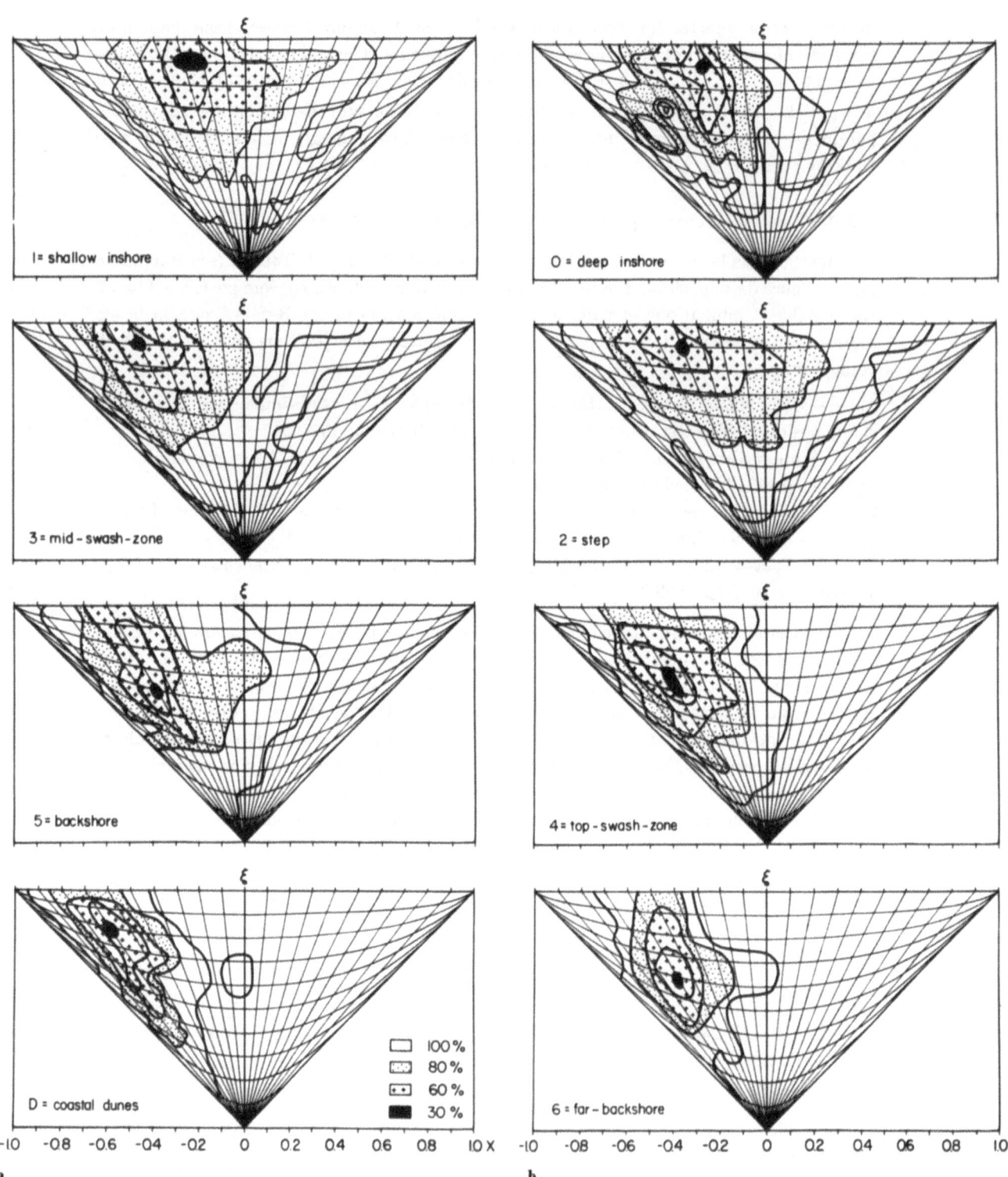

Fig. 5. Contoured shape positions of the eight cross-shore environments from segment A shown in the hyperbolic shape triangles

that the backshore and the coastal dune sediments are significantly different points to the fact that the shape of the samples changes gradually from one environment to the other. The far-backshore sediments still show the shape of the backshore but already bear the finger prints of the selective sorting process of the wind.

The top-swash-zone and the backshore sediments are highly correlated, and the sedimentological explanation is that most of the backshore sediments are actually deposited as top-swash-zone sediments during stormy conditions. Thus the selective sorting processes which are responsible for the deposition of the backshore and top-swash-zone sediments are the same. The phenomenon that was observed before is repeated. The top swash-zone sediments are very similar to the backshore sediments but are significantly different from the far-backshore and the coastal dune sediments. There is a gradual and significant change in the shape characteristics of the grain-size distribution, starting with the top swash-zone and ending with the coastal dune sediments.

The relation between the top-swash-zone sediments and the seaward neighbour, mid-swash-zone environment is different. These two super samples have totally different shape position distributions. It looks as if the subaqueous and subaerial sediments make up two very different distributional shape groups.

The relationship between the subaqueous environments is not simple. The test criterion for the deep inshore sediments shows that there is no significant correlation with the other environments from this group. The other super-samples are significantly different one from the other, though the relationship between the step and the mid-swash-zone sediments is marginal.

However, we have to recall that the backshore and top-swash-zone sediments are waterlain sediments and not aeolian. This means that the selective sorting processes in the swash-zone acting on an available wide range of grain-sizes produce backshore sediments with a skewed shape as source material for the aeolian activity. This shape is changed by the aeolian processes, usually leading towards lower ϱ values, i.e., positively skewed in the commonly used terminology. When the source sediments have a narrow range of grain sizes with a different initial distributional shape, the aeolian end-products may evolve in another way (Barndorff-Nielsen and Christiansen [32] and Barndorff-Nielsen and Sørensen [33]). This can explain the fact that some studies, summarized in McKee [34], reported negative skewness (in the phi scale).

5.2.2 Shape positions (χ, ξ) distributions - shape triangle contour maps

Figure 5 shows the contoured distribution maps of the shape positions in the domain of the hyperbolic shape triangle. The contours were drawn to enclose 100%, 80%, 60%, and 30% of the samples. The location with the highest density represents the mode region of the joint distribution of χ and ξ. These contour maps demonstrate the distribution of the shape positions in each environment and support qualitatively our understanding, as discussed above, of the relation between them. Although the variability and the overlap between the populations seem to be large, we have to remember that the shape positions have a probabilistic nature (Hartmann [15], Barndorff-Nielsen et al. [20]). In the previous subsection it was demonstrated that most of these populations are significantly different, and it was inferred that there is a dynamical connection between the sequence of the environments.

Whereas the subaqueous group has contours covering a wide range of shape positions, the subaerial group is restricted to the left half of the shape triangle. This means that the

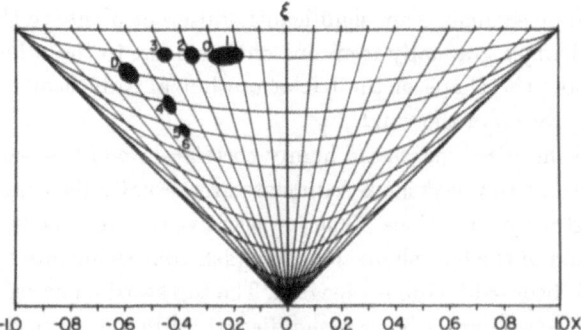

Fig. 6. Modal regions of the distributions of shape positions of the eight cross-shore environments in the hyperbolic shape triangle

shape of a subaqueous sediment sample can be skewed either way or be symmetrical. The subaerial sediments always have a steep, coarse tail and a relatively flat, fine tail.

In the subaqueous group the shallow inshore sediments have the biggest variability of shape positions. They have more symmetrical and fine skewed grain-size distributions than the other super-samples. The contours of the deep inshore are located near the center of the shape triangle and therefore are not significantly different from the other subaqueous super-samples. The contoured shape positions of the mid-swash zone and the step have a common boundary on the left of the triangle but the step samples are more symmetrical, with a higher percentage of fine-skewed distributions.

In Figure 6, where only the modal regions of the different environments are plotted, a general pattern of the entire cross-shore sequence of the environments has emerged. There is also here a clear trend from the inshore, via the swash-zone to the backshore, ending up with the aeolian sediments. The inshore sediments are the most symmetrical and the aeolian sediments most asymmetrical. The modes of the subaqueous sediments lie on the same 'hammock' curve which indicates their high steepness, whereas the subaerial sediments are less peaked and lie on the same ϱ line (-0.6 to -0.7ϱ). It was already demonstrated (Hartmann and Christiansen [35], Hartmann [15]) that the tilt parameter ϱ is best suited for differentiating very closely connected subenvironments in the shape triangle. In conclusion, the sediments differ not only in the tilt (or asymmetry) but also in the steepness of their distributions. This difference is environmentally significant as we are analysing the populations and not the single samples.

6 Discussion: cross-shore grain-size selective sorting

Parameters of particle size distributions are routinely used in the geological literature and there are numerous publications in the literature about this topic. However, there is no general agreement as to how the parameters should be estimated or how the distributions should be described (Inman [36], Jones [37], Dapples [38], Le Roy [39], Christiansen et al. [40], Christiansen and Hartmann [41], Hartmann [15], [42]), and there is, in fact, considerable controversy in the field. As was discussed before, with the data of a single sample one must always raise doubt about the information content and the information/noise ratio and thus the accuracy of its statistical parameters.

As can be expected from a natural probabilistic model, it was demonstrated that the sediment sorting processes do not create sharply defined environments but rather populations with big overlaps in the distributions of their significant attributes. In spite of these un-clear boundaries, each of the environments is statistically unique and bears its specific environmental fingerprints.

The estimated textural parameters should be the independent variables in the environ-mental discrimination procedures. In the present study it became clear, in agreement with many other studies (e.g. Vincent [43], Hartmann [15], Barndorff-Nielsen et al. [20]), that none of the parameters and their derivatives fulfil the necessary conditions for most types of multivariate data analysis. However, as the probabilistic nature of the textural attributes was exposed, it is now possible to make a parametric feature extraction for the unimodal distributions and study it in a more proper way.

The present study has provided significant information about the relationship between grain size distributions and cross-shore environments. Specific processes control the sedi-ments in the sequence of eight different beach subenvironments and are reflected in the shape of their grain size distributions.

Many of the sedimentological studies involving grain-size analysis relate selected com-binations of statistical granulometric parameters to various depositional environments. The common opinion is that these attributes reveal environmentally specific information which is hidden in the distributions. Passega's [44] approach was to distinguish between modes of transport which actually reflect the dynamical processes rather than between environments. However, in many sedimentological environments there is more than one mode of transport (Visher [45]). Various techniques were developed to assign certain parts of a grain size distribution to a certain mode of transport. Some of the results are still controversial and were shown by Kennedy et al. [46] as not very reliable. However, many studies reported connections between segments on a log-normal cumulative probability paper, and modes of transportation (Visher [45], Kolmer [47], Reed et al. [23], Middleton [48], Bryant [8]). Christiansen et al. [40] have shown that the origin of these segments is just a quantification error caused by not using the proper probability model. This paradox can be solved by assuming that some properties of the hyperbolic distribution, like the slopes and their inflection with the modal region, are reflected in the segmented line on a normal probability paper. By using the data and results from this study, it can be concluded that the hyperbolic grain size distribution contains more information on sedimentary processes than we can explain at present.

Because of the complexity of the processes involved in sediment transport, Barndorff-Nielsen and Christiansen [32] in their erosion-deposition model and the newly presented enlarged model by Barndorff-Nielsen and Sørensen [33] simplify the dynamical behaviour of the transport of grains in relation to fluid mechanics. The models generalize the entrainr-ment of the sedimentary grains and do not consider how they move, or from which mode of transport they are deposited.

The complexity of the processes involved is summarized in Bridge [49] (p. 1110) "... in a ... turbulent flow, individual grains of a given size may well travel in more than one transport mode ...". There are no reports in the literature to suggest that a single 'segment' could be observed and that this observation could be directly related to a single mode of transport. Hartmann [15], [50] deals with several hundreds of samples which were trapped during movement in the mid-swash-zone. Are they therefore the outcome of one transpor-tation mode? The results show that this hypothesis is not correct and even erroneous. What becomes clear from the various studies is that there are ranges of grain-sizes which 'prefer' to rest under certain conditions and are not stable in other ones. Liu and Zarillo [51] summarize results presented by Ingle [6], Duane [52] and Bridge [49] and conclude

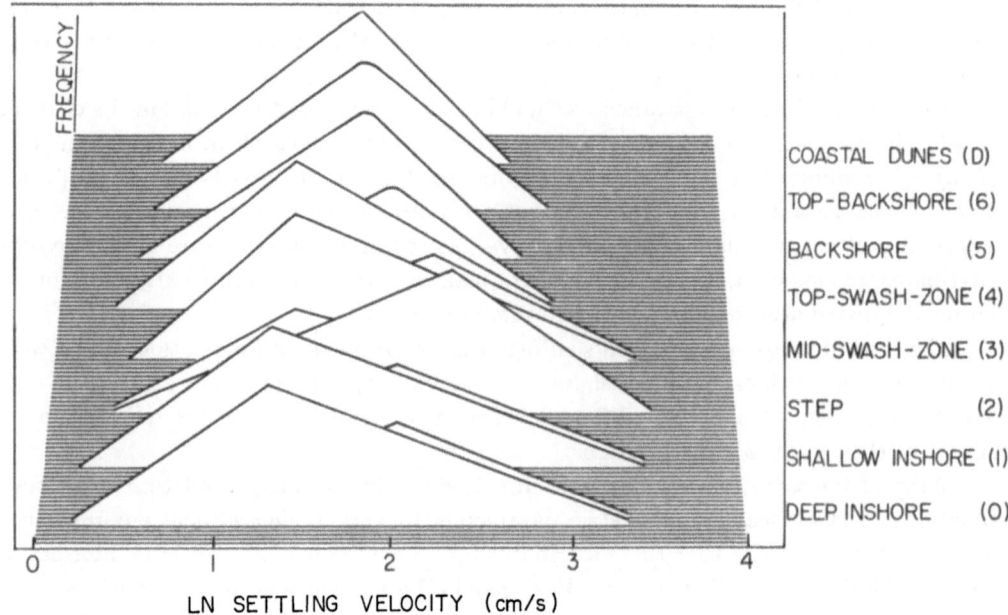

Fig. 7. A diagram showing the tails cutoffs and the modal points of the eight grain-size super-samples of the beach environments

that different sediment sizes respond differently to the same hydrodynamics. Figure 7 demonstrates the cutoffs in the fine and coarse tail and the modal point of the eight grain size populations studied here. The truncation at the fine end is related to the cutoff caused by the predominant winnowing of suspended grains. The coarse-end cutoff is related to the spatial velocity of the creeping and rolling grains. Hartmann [15], [53] demonstrates that in each of the various cross-shore depositional environments, the typical (modal) grain size can be related to the predominant mode(s) of transport. Thus the relation between modes of transport in a given environment dictates the three points which form the shape of a distribution: the modal point and the two limiting points at the tails.

Although there are significant differences between the morphology of the beach profiles and the prevailing wave climate at the two seasons studied, the grain-size analysis does not reflect these changes. The results reported here derive from the samples of the autumn survey. The textural variability between the seasons is very small, and the spring survey populations are very similar, within their variability range, to the other season (Hartmann [15], [53]). This shows that the studied populations reflect the sum of the long-term processes and may suggest that the sedimentary system along the southern coast of Isreal has reached a stage of textural equilibrium.

One also has to consider the fact that the spatial and temporal variability of the energy factors, the gustiness or bursting rate, can mimic varying modes of transport. The critical traction and suspension velocities for particles in the foreshore depend on the flow depths which can vary greatly from a few millimeters to several centimeters, depending on the characteristics of the nearby breaking wave (Bryant [8]). Investigations to trace the different modes of transport require a very high level of resolution, and the dynamical measurements and morphological characterization should be kept at the same level. At a lower level of resolution one can characterize the different environments by using a probabilistic approach, as was done in this paper.

The selective size-sorting processes between the populations sort out and differentiate very typical grain size populations. These processes are defined here as sorting differentiation processes. The approach presented in this study is qualitative with regard to dynamics. It provides no information regarding the level of the energy involved.

Acknowledgements

I thank Dr. R. Goldberry from the Department of Geology at Ben Gurion University, Beer Sheva, Israel, for his support during the fieldwork and the laboratory work. Many thanks are due to Dr. Ch. Christiansen from Aarhus University, Denmark, for his active support during the data analysis and interpretation phase. Thanks are also due to Prof. O. Barndorff-Nielsen and Dr. P. Blæsild for many fruitful and encouraging discussions. Prof. D. Bowman, Prof. Barndorff-Nielsen, Dr. Christiansen and an anonymous reviewer helped by critically reviewing the manuscript.

References

[1] Evans, O. F.: Sorting and transportation of material in the swash and backwash. J. Sed. Petrol. **9**, 28—31 (1939).

[2] Miller, R. L., Zeigler, J. M.: A model relating dynamics and sediment pattern in equilibrium in the region of shoaling waves, breaker zone, and foreshore. J. Geol. **66**, 417—441 (1958).

[3] Seibold, E.: Geological investigations of nearshore sand transport — examples of methods and problems from the Baltic and North seas. Progr. Oceanogr. **1**, 3—70 (1963).

[4] Schiffman, A.: Energy measurements in the swash-surf zone. Limmol. Oceanog. **10**, 255—260 (1965).

[5] Fox, W. T., Ladd, J. W., Martin, M. K.: A profile of the four moment measures perpendicular to a shore line, South Haven, Michigan. J. Sed. Petrol. **36**, 1126—1130 (1966).

[6] Ingle, J. C.: The movement of beach sand. Amsterdam: Elsevier 1966.

[7] Greenwood, B., Davidson-Arnott, R. G. D.: Textural variation in subenvironments of the shallow-water wave zone, Kouchibouguac Bay, New Brunswick. Canad. J. Earth Sc. **9**, 679 to 688 (1972).

[8] Bryant, E. A.: Relationship between foreshore sediment settling velocity and breaker wave hydrodinamics, eastern Australian beaches. Ph. D. thesis (unpub.), Macquarie University, Australia 1977.

[9] Komar, P. D.: Beach processes and sedimentation. Englewood Cliffs, N. J.: Prentice-Hall 1976.

[10] Bluck, B. J.: Sedimentation of beach gravels: examples from south Wales. J. Sed. Petrol. **37**, 128—156 (1967).

[11] Pranzini, E.: Random changes in beach sand grain-size parameters. Boll. Soc. Geol. It. **102**, 177—189 (1983).

[12] Sedimentation Seminar: Comparison of methods of size analysis for sands of the Amazon-Solimoes Rivers, Brazil and Peru. Sedimentology **28**, 123—128 (1981).

[13] Gibbs, R. J.: A settling tube system for sand-size analysis. J. Sed. Petrol. **44**, 583—588 (1974).

[14] Geldof, H. J., Slot, R. E.: Settling tube analysis of sand. Internal Report, Dep. Civil Eng. Delft University **4—79**, 1—31 (1979).

[15] Hartmann, D.: Coastal sands of the southern part of the Meditarranean coast of Israel: reflections of dynamic sorting processes and environmental discrimination. Ph. D. Thesis (unpub.), Aarhus, Denmark 1988a.

[16] Kachholz, K. D.: Statistische Bearbeitung von Probendaten aus Vorstrandbereichen sandiger Brandungsküsten mit verschiedener Intensität der Energieumwandlung. Ph. D. Thesis (unpub.), Kiel, Germany 1982.

[17] Krumbein, W. C., Graybill, F. A.: An introduction to statistical models in geology. McGraw-Hill, 1965.

[18] Russell, R. J.: Where most grains of very coarse sand and fine gravel are deposited. Sedimentology **11**, 31—38 (1968).

[19] Shea, J. H.: Deficiencies of clastic particles of certain sizes. J. Sed. Petrol. **44**, 985—1003 (1974).

[20] Barndorff-Nielsen, O. E., Christiansen, C., Hartmann, D.: Distributional shape triangles with some applications in sedimentology. Acta Mechanica [Suppl. 2]; 37—47 (1991).

[21] Mayo, W.: A computer program for calculating statistical parameters of grain-size distributions derived from various analytical methods. B. M. R. Rec. (unpub.) **140**, 1—23 (1972).

[22] Goldbery, R., Tehori, O.: SEDPAK — a comprehensive operational system and data-processing package in APPLESOFT BASIC for a settling tube, sediment analyzer. Comput. Geosci. **13**, 565—585 (1987).

[23] Reed, W. E., LeFever, R., Moir, G. J.: Depositional environment interpretation from settling-velocity (Psi) distributions. Geol. Soc. Am. Bull. **86**, 1321—1328 (1975).

[24] May, J. P.: Chi (χ): a proposed standard parameter for settling tube analysis of sediments. J. Sed. Petrol. **51**, 607—610 (1981).

[25] Bryant, E. A.: Sample site variation in settling velocity distribution subpopulations using curve dissection analysis. Sedimentology **33**, 767—775 (1986).

[26] Christansen, C., Hartmann, D.: SAHARA: a package of PC programs for estimating both log-hyperbolic grain-size parameters and standard moments. Comput. Geosci. **14**, 557—625 (1988a).

[27] Jensen, J. L.: Maximum likelihood estimation of the hyperbolic parameters from grouped observations. Comput. Geosci. **14**, 1—38 (1988).

[28] Hartmann, D., Christiansen, C.: The hyperbolic shape triangle as a tool for discrimination of populations of sediment samples from closely connected origin. Math. Geology, submitted (1990).

[29] Barndorff-Nielsen, O. E., Blæsild, P., Jensen, J. L., Sørensen, M.: The fascination of sand. In: A celebration of statistics. Centenary Volume of the International Statistical Institute. New York: Springer 1985.

[30] Barndorff-Nielsen, O. E., Blæsild, P.: Hyperbolic distributions and ramifications: contributions to theory and applications. Statistical distribution in scientific work. Dordrecth: D. Reidel Publ. Co. 1981.

[31] Bagnold, R. A., Barndorff-Nielsen, O.: The pattern of natural size distribution. Sedimentology **27**, 199—207 (1980).

[32] Barndorff-Nielsen, O. E., Christiansen, C.: Erosion, deposition and size distributions of sand. Proc. R. Soc. Lond. **A 417**, 335—352 (1988).

[33] Barndorff-Nielsen, O. E., Sørensen, M.: On the temporal-spatial variation of sediment size distributions. Acta Mechanica [Suppl. 2]; 23—35 (1991).

[34] McKee, E. D.: A study of global sand seas. Geological survey professional pap., Washington 1979.

[35] Hartmann, D., Christiansen, C.: Settling-velocity distributions and sorting processes on a longitudinal dune: a case study. Earth. Surf. Proc. Landforms **13**, 649—656 (1988).

[36] Inman, D. L.: Measures for describing the size distribution of sediments. J. Sed. Petrol. **22**, 125 to 145 (1952).

[37] Jones, T. A.: Comparisons of the descriptions of sediment grain-size distributions. J. Sed. Petrol. **40**, 1204—1215 (1970).

[38] Dapples, E. C.: Laws of distribution applied to sand sizes. G. S. A. Memoir **142**, 37—61 (1975).

[39] LeRoy, S. D.: Grain-size and moment measures: a new look at Karl Person's ideas on distributions. J. Sed. Petrol. **51**, 625—630 (1981).

[40] Christiansen, C., Blæsild, P., Dalsgaard, K.: Re-interpreting 'segmented' grain-size corves. Geol. Mag. **121**, 47—51 (1984).

[41] Christiansen, C., Hartmann, D.: On using the log-hyperbolic distribution to describe the textural characteristics of aeolian sediments-Discussion. J. Sed. Petrol. **58**, 159—160 (1988b).

[42] Hartmann, D.: The goodness-of-fit to ideal Gauss and Rosin distribution: a new grain-size parameter-Discussion. J. Sed. Petrol. **58**, 913—917 (1988b).

[43] Vincent, P.: Differentiation of modern beach and coastal dune sands — a logistic regression approach using the parameters of the hyperbolic function. Sed. Geol. **49**, 167—176 (1986).

[44] Passega, R.: Grain size representation by CM patterns as a geological tool. J. Sed. Petrol. **34**, 830—847 (1964).

[45] Visher, G. S.: Grain-size distribution and depositional processes. J. Sed. Petrol. **39**, 1074—1106 (1969).

[46] Kennedy, S. K., Ehrlich, R., Kana, T. W.: The non-normal distribution of intermittent suspension sediments below breaking waves. J. Sed. Petrol. **51**, 1103—1108 (1981).

[47] Kolmer, J. R.: A wave tank analysis of the beach foreshore grain size distribution. J. Sed. Petrol. **43**, 200—204 (1973).

[48] Middleton, G. V.: Hydraulic interpretation of size distributions. J. Geol. **84**, 405—426 (1976).

[49] Bridge, J. S.: Hydraulic interpretation of grain-size distributions using a physical model for bedload transport. J. Sed. Petrol. **51**, 1109—1124 (1981).

[50] Hartmann, D.: Swash-zone sediment dynamics in the light of the population concept and the hyperbolic model. (In preparation) (1991).

[51] Liu, J. T., Zarillo, G. A.: Distribution of grain sizes across a transgressive shoreface. Marine Geol. **87**, 121—136 (1989).

[52] Duane, D. B.: Synoptic observations of sand movement. Proc. Coastal Eng. Conf. 12th (1970).

[53] Hartmann, D.: Cross-shore sorting differentiation processes dominating a beach-dune system. 13th International Sedimentological Congress (abs.), 215 (1990).

Author's address: D. Hartmann, Ph. D., National Institute of Oceanography, Israel Oceanographic and Limnological Research, Tel Shikmona, P. O. Box 8030, Haifa 31080, Israel

Acta Mechanica (1991) [Suppl] 2: 65—75

Roughness element effect on local and universal saltation transport

J. D. Iversen, W. P. Wang, Ames, Iowa, USA, **K. R. Rasmussen, H. E. Mikkelsen,** Aarhus, Denmark, and **R. N. Leach,** Mountain View, California, USA

Summary. Experimental results are presented which illustrate the effects of permanent surface obstructions on saltation phenomena. It is shown that the topographic drift geometry and the dimensionless erosion rates of windward erosion associated with cylindrical obstacles are strong functions of the cylinder aspect ratio. For short cylinders, there is also significant erosion taking place in the far wake. These two erosional areas develop due to different sets of separation vortex systems. For multi-element roughness arrays, sparse array data are presented which illustrate the increase of threshold friction speed with element frontal area density and roughness element drag coefficient.

1 Introduction

The aerodynamic characteristics of boundary layer surface protrusions and their effects on aeolian transport are problems of continuing significance. Characterization of surface roughness as to its effects on the boundary layer is a difficult problem with a long history of study [1], [2], [3]. The effects of roughness on saltation phenomena have also been studied but with little success at generalization [4], [5], [6].

In this paper, we discuss recent experiments concerning the effects of both single roughness elements and distributed roughness on aeolian phenomena. The effects of single boundary layer protrusions on aeolian phenomena are important in terms of drift control, design of structures to minimize drift problems, interpretation of aeolian geomorphology, and the study of parametric effects in modeling such phenomena at small scale [7].

Studies of the aerodynamic effects of single elements have been performed, e.g., by a number of investigators [8], [9], [10], [11], [12], [13]. The effects of single elements on aeolian drift topography have also been studied, but not as extensively [14], [15], [16], [17], [18], [19]. We show here that for prismatic shapes and cylinders, significant changes in drift topography result with changes in aspect ratio (defined here as the ratio of element height to lateral breadth). Data are also presented to illustrate the effects of multiple element geometry and density on aeolian saltation threshold.

2 Experimental results — single elements

The flow pattern, and therefore the drift topography, varies considerably with change in obstacle shape and aspect ratio. These variations are clearly exemplified by the differences between the rectangular prism (with one face normal to the wind direction) and the circular cylinder as shown in Figs. 1, 2. In Fig. 1 the drift pattern associated with a rectangular prism is illustrated. There is evidence of both primary and secondary vortices on the windward erosional pattern in Fig. 1. Although Froude numbers and particle characteristics are different in the two cases, the primary differences are due to effects of geometry.

In Fig. 2, the drift topographies adjacent to circular cylinders are illustrated. For the prism and for the taller cylinders (2a, b), a strong "horseshoe" type vortex exists windward of the obstacle and the windward eroded moat or scour area is the most prominent feature.

The drift topographic characteristics in the obstacle wakes of the prism and the cylinders, however, are quite different, mostly due to differences in the manner in which flow separates from the top surfaces. There is little vorticity shed from the upper surface of the prism [11], but there can be significant vorticity shed from a horizontal surface which has a windward edge not parallel or perpendicular to the wind direction [20]. Since the cylinder top is such a surface, sufficiently significant vorticity is shed to not only cause an evident difference in the near wake as seen in Figs. 1, 2, but also for a considerable distance downwind, where enhanced erosion occurs in the circular cylinder wake but not in that of the prism.

The type of this enhanced erosion in the far wake for the circular cylinder depends on the aspect ratio (diameter to exposed height ratio). In Fig. 2a a cylinder of aspect ratio 2.77 exhibits a single erosional streak which extends for many diameters downwind. This streak starts just downwind of twin diagonal deposit ridges in the near wake which are formed in the separation region just downwind of the cylinder. There are some similarities here with the cube wake topography. For shorter cylinders, as in Fig. 2b (aspect ratio = 1.29), there are two erosional streaks in the far wake, corresponding to the horizontal vortices shed from either side of the cylinder top. As the height/diameter ratio decreases still further, the erosional capability of these vortices increases in the parallel far downwind wake, illustrated in Fig. 2c. There are two other significant differences in the drift topography for the shortest cylinder as compared to the taller cylinders shown in Fig. 2a, b. The near wake deposit region is quite different and the ridges have disappeared. Perhaps even more striking is the fact that the upwind erosional scoured region has disappeared. The portion of the

Fig. 1. Topographic drift pattern for $8 \times 8 \times 16$ cm prism. Flow left to right. Iowa State University wind tunnel. Sand particles 590 μm average diameter. Froude number $u_*^2/gD = 0.44$ (D is lateral width = 16 cm). Reynolds number $U_h D/\nu = 75000$

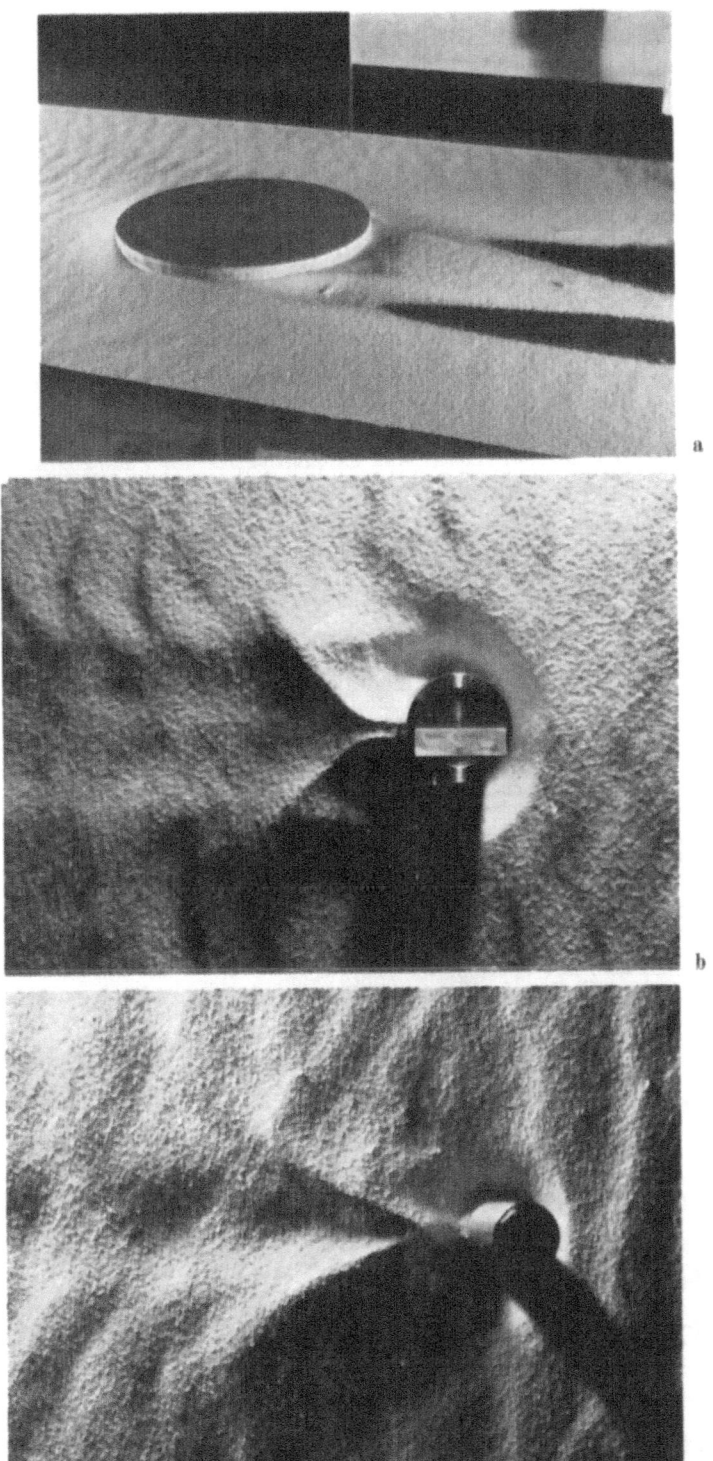

Fig. 2. Topographic drift patterns for cylindrical elements. **a** Drift pattern for cylinder of exposed height 8.3 cm and diameter 3 cm. Aarhus University wind tunnel. Sand particles 220 μm average diameter. Froude number $u_*^2/gD = 0.66$. Reynolds number $U_h D/\nu = 13,000$. **b** Drift pattern for cylinder of exposed height 3.2 cm and diameter 5 cm. Aarhus University wind tunnel. Froude number $u_*^2/gD = 0.40$. Reynolds number $U_h D/\nu = 19,000$. **c** Drift pattern for cylinder of exposed height 3 cm and diameter 39 cm. Iowa State University wind tunnel. Froude number $u_*^2/gD = 0.18$. Reynolds number $U_h D/\nu = 190,000$

horseshoe vortex in the upwind portion of the separated flow region is apparently too weak at the start of motion to prevent deposition of sand immediately upwind of the short, broad cylinder. The vortices shed from the cylinder top surface, however, have a powerful erosive capability for many diameters downwind.

To determine the time-dependent detailed evolution of the drift topography adjacent to a series of circular cylindrical models, a moiré photographic technique was utilized [19], [21]. A photograph of a diffraction grating is optically projected onto the horizontal surface of interest at a predetermined angle to the vertical. A camera is placed above the surface with its focal plane horizontal. Topographic contour lines are obtained by superposition of two transparent images, one of a plane white surface and the other of the evolving drift surface. In order to obtain volumetric data, the moire contours were digitized and volume removed from the upwind erosional region was calculated for each photograph, obtaining a time-dependent history of eroded volume.

The windward erosional scour geometry is a strong function of aspect ratio as shown in Figs. 3a, b. The scour depth increases with aspect ratio as shown in Fig. 3a. The data envelope as represented by the curved line may be close to the equilibrium value of scour depth, even though the experimental times were not sufficiently large for equilibrium to be reached. Data values below the line are for experiments of too short duration for equili-

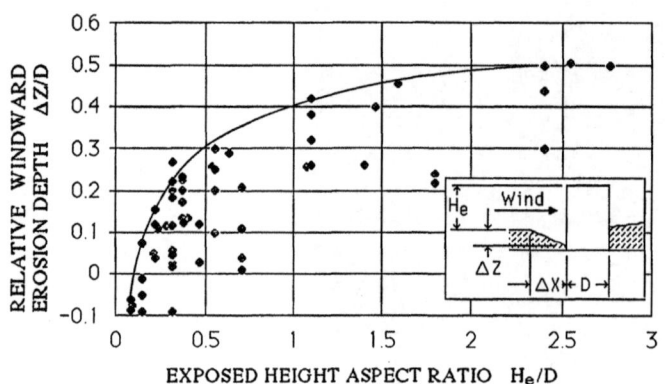

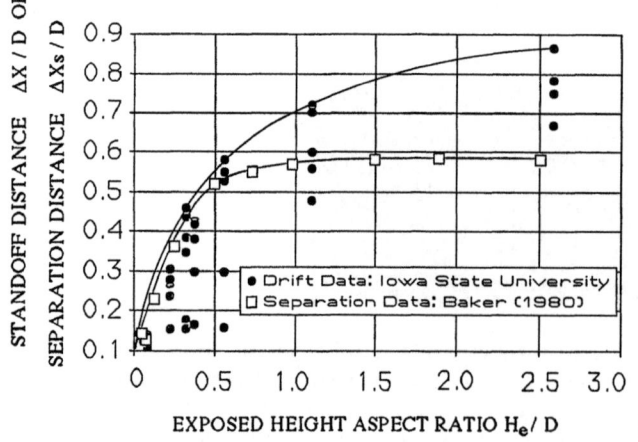

Fig. 3. Upwind erosional dimensions for circular cylinders as functions of aspect ratio. Iowa State University wind tunnel. **a** Upwind scour depth versus aspect ratio. For aspect ratios less than 0.08, the depth is negative (deposition rather than erosion). **b** Standoff distance versus aspect ratio

brium to be approached. For values of aspect ratio less than about 0.1 (the short, broad cylinder), the "scour" depth is negative, meaning that deposition occurs upwind (as in Fig. 2c).

The upwind extent of the windward scour zone, or "standoff" distance x, is represented by the data in Fig. 3b. Also illustrated in this figure is the separation distance as measured by Baker [13] (i.e., the point on windward centerline of primary separation as measured by oil-flow visualization). There is obviously a strong correlation between his separation distance data and the present upwind scour distance data, for aspect ratios less than 0.5. For larger aspect ratios with the resultant stronger horseshoe vortex, the drift surface topography may be too far different from the flat surface in Baker's experiments.

It was possible, from the digitized contour data, to obtain the volume of the windward scoured region as a function of time. These data, for seven different aspect ratios, are shown in Fig. 4a, b. For most of these experiments the volume rate of erosion from the windward region was constant with time, as illustrated by the straight lines in Fig. 4a. It was necessary to terminate the experiments before the scoured volume reached the test section floor.

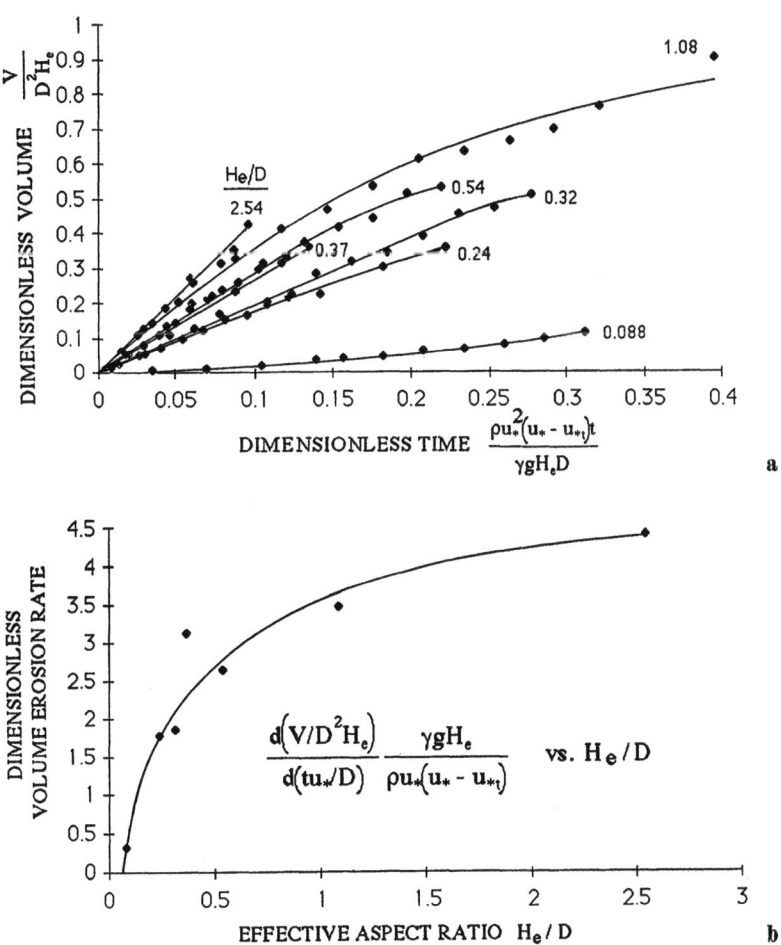

Fig. 4. Upwind erosional volume for cylinders of various aspect ratios. Data were obtained from Moiré photographic technique [19]. Iowa State University wind tunnel (ϱ is air density, γ is particle bulk density, t is time). **a** Dimensionless volume versus dimensionless time. **b** Dimensionless erosion rate versus aspect ratio. These data points represent the slopes of the curves for small time in Fig. 4a

For three of the intermediate aspect ratios, however, the volume rate has begun to decrease towards the end of the experiment and it appears that the approach to equilibrium scour geometry has begun.

The data are plotted in dimensionless form in Fig. 4a to show that the relative scour rate increases with increase in cylinder aspect ratio. The slopes of these curves for small time t in Fig. 4a are plotted against aspect ratio in Fig. 4b. As with the geometry changes noted in Fig. 3, the greatest changes with aspect ratio occur for aspect ratios between zero and one. Also, the erosion rate is negative for aspect ratios less than about 0.1, similar to the data shown in Fig. 3a. No negative results are shown in Fig. 4a, however, since the moiré equipment was not in place when experiments with deposition were obtained.

The drift geometry can be written in functional form as follows:

$$\text{DRIFT GEOMETRY} = f\left(\frac{u_*}{u_{*_t}}, \frac{u_*^2}{gD}, \frac{z_0}{D}, \frac{H_e}{D}, \frac{\varrho}{\gamma}, \frac{u_* D}{\nu}, \frac{u_*^3}{gv}, \frac{H_e}{\delta}, \frac{U_F}{u_{*_t}}, \frac{tu_*}{D}\right)$$

where z_0 is the aerodynamic roughness height if nonerodible roughness elements are present, ϱ and γ are fluid and particle densities, ν is kinematic viscosity, ϱ is the boundary layer thickness, t is time and U_F is particle terminal speed. In the experiments performed thus far, we have systematically investigated only the effects of changes in aspect ratio H_e/D and time tu_*/D. It is expected that there are also at least subtle effects of all the parameters listed in the equation. The values of Froude and Reynolds numbers in the Aarhus and ISU wind tunnels are sufficiently close that large differences in drift geometry are not expected. In order to find the effects of all of the parameters, it will be necessary to perform many more experiments.

The values of u_* and z_0 for the Aarhus and ISU wind tunnel tests were determined from mean velocity profiles taken sufficiently far from the models. Wind speeds in the ISU tunnel were higher ($u_* = 83$ cm/s) than in the Aarhus tunnel ($44 < u_* < 50$ cm/s) because of the larger particle diameter used in the ISU tests.

3 Experimental results — multiple elements

A second series of experiments involved multi-element roughness arrays. The presence of a multi-element set of non-erodible roughness elements creates two primary effects on the saltation process, i.e., (1) it increases the value of threshold shear stress and (2) it decreases the mass transport rate for a given value of shear stress [4], [5], [22]. Both of these effects occur because the portion of the mean surface shear stress due to the roughness elements themselves (primarily due to pressure drag) is greater than the mean value. The average shear stress on the erodible particle bed portion of the surface is thus less than the mean value. The mean value of shear stress at initiation of movement is thus higher than the normal threshold value.

The equivalent roughness height (and therefore the mean surface friction speed) is a complex function of the size, shape and distribution of the roughness elements. Values of the relative roughness height Z_0/H were determined from experimental velocity profile data for a particular kind of roughness array (rectangular blocks of variable height H of lateral extent twice the windward extent and for which the ratio of plan area to total horizontal area is 0.0366 [23]. An approximate fit to these data is

$$\frac{Z_0}{H} = 5 \text{ (Frontal Area/Area)}^{1.4}, \ (0{,}006 \leq \text{Frontal Area/Area} \leq 0{,}06)$$

This equation is closely related to Lettau's [24]. The values of Z_0 were determined from regressional fits of the boundary layer velocity profile (determined from pitot probe traverses) to Prandtl's boundary layer equation

$$\frac{u}{u_*} = \frac{1}{k} \ln \left(\frac{Z - D}{Z_0} \right)$$

where D is assumed in this case to be the average height of the surface with respect to the measurement plane. The values of friction speed ratio u_*/U obtained from these profiles and others for block arrays of the same configuration, obtained from three different wind tunnels [23], [25], all agree well with Schlichting's Moody Diagram [1].

Equilibrium drift topographies for two different exposures of the block arrays are shown in Fig. 5 with increasing block exposure illustrated in successive photographs [26]. For the lowest exposure height (Fig. 5a), the dominant feature is the ridged wake deposition pattern. As the exposure height increases, the windward erosional volume

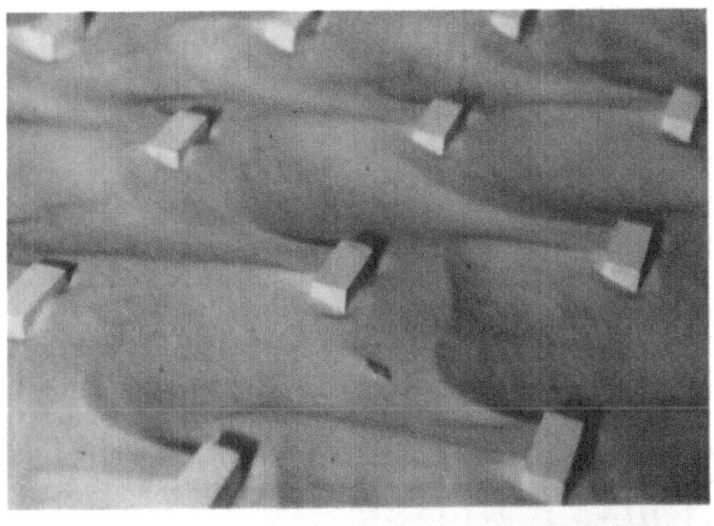

Fig. 5. Photographs of equilibrium drift topography for block arrays of different exposure heights· The blocks are of horizontal dimensions 1.91×3.81 cm [26]. **a** Small exposure height. Flow right to left. Froude number $u_*^2/gD = 0.37$. $Z_0 \cong 0.007$ cm. **b** Large exposure height. Flow left to right. Froude number $u_*^2/gD = 1.13$. $Z_0 \cong 0.3$ cm

increases and the leeward double ridges coalesce into a forked pattern. The blocks in these arrays are sufficiently far apart that the flow and deposition patterns associated with each block seem to be relatively unaffected by upwind blocks. If these blocks were arrayed in tighter patterns, however, the flow and deposition patterns would change significantly.

The threshold friction speed for 145 m sand particles for this block array was determined experimentally for several exposure heights at both atmospheric and at reduced pressure [26]. The data for low pressure are plotted in Fig. 6a. Plotted in Fig. 6b are data for arrays of circular cylinders [6] of various sizes, exposure heights and array densities and for hemispheres [28] for various array densities and sphere sizes. The data in Figs. 6a, b are replotted in Fig. 6c. Gillette and Stockton [28] indicate that the appropriate parameter for determining the effect of roughness is the element frontal area per unit floor area. It

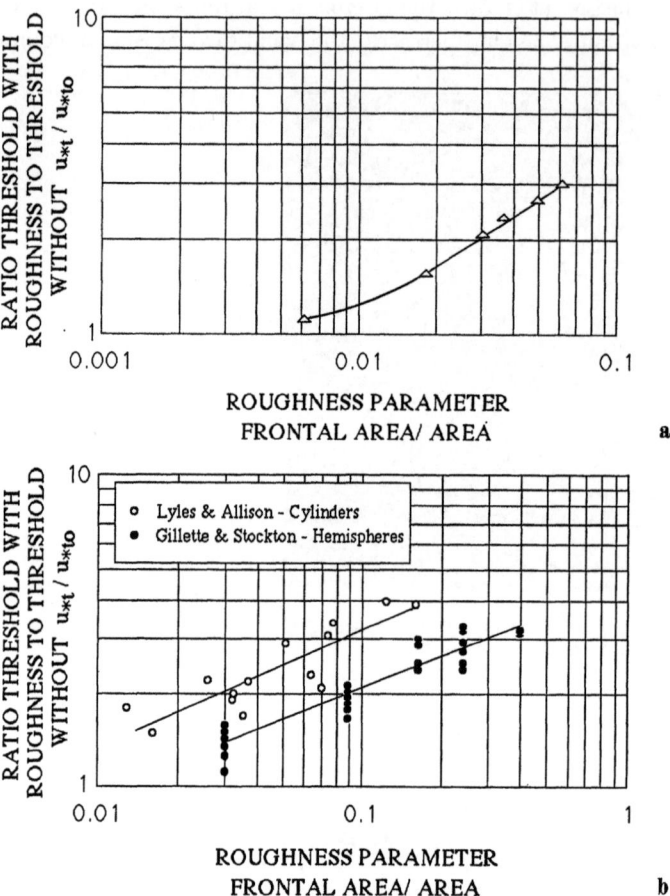

Fig. 6. Ratio of threshold friction speed in the presence of uniform roughness elements as a function of element shape and array density. Data for rectangular blocks, circular cylinders, and hemispheres. Values of threshold friction speed were determined for surfaces of a stable equilibrium shape (rather than a smoothed, level surface). **a** Rectangular blocks [26]. **b** Cylinders [6] and hemispheres [28]

$$\frac{u_{*_t}}{u_{*_{t_0}}} = 7.56 \left(\frac{\text{Frontal Area}}{\text{Floor Area}}\right)^{0.374}, \text{ cylinders}$$

$$\frac{u_{*_t}}{u_{*_{t_0}}} = 4.91 \left(\frac{\text{Frontal Area}}{\text{Floor Area}}\right)^{0.335}, \text{ hemispheres.}$$

$$\frac{u_{*_t}}{u_{*_{t_0}}} = 6.33 \left(\frac{C_D \times \text{Frontal Area}}{\text{Floor Area}}\right)^{0.355}, \text{ c Blocks, cylinders, and hemispheres.}$$

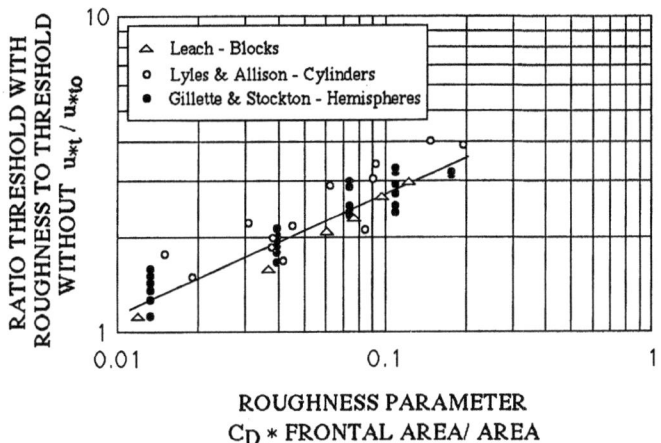

ROUGHNESS PARAMETER
$C_D *$ FRONTAL AREA/ AREA   **Fig. 6c**

appears that the shape of the roughness element is also important. The drag coefficients used in Fig. 6c are 2 for the rectangular blocks [29], 1.2 for the circular cylinders and 0.45 for the hemispheres [30]. As exposure height or array density increases, the value of friction speed at threshold increases. For a given kind of roughness array, there is a value of array element density (frontal area per floor area approximately = 0.1) which results in a maximum value of equivalent roughness height [3]. Gillette and Stockton's data show that the protection afforded erodible particles lying on the surface between the nonerodible roughness elements continues to increase as array element density increases beyond the critical roughness density value of 0.1, as would be expected.

4 Conclusions

The flow pattern associated with a single roughness element which protrudes into the boundary layer from a flat surface is not known with any detailed precision. Although a number of experimental measurements have been made [11], such flows except at very low Reynolds number are quite unsteady, and thus measurements, e.g., of wake vortex locations and strengths which are necessarily performed by time-averaging techniques are probably not very indicative of the instantaneous flow structure. When aeolian transport is added to the flow, the situation becomes even more complex. The drift topography does not exactly mimic the surface shear stress pattern which exists in the absence of particle flow because (1) aeolian motion itself affects the local shear stress as well as the flow pattern past the obstacle, (2) as drift topography evolves, the local stress pattern will change accordingly, and (3) there is a downstream effect, particularly for large Froude numbers u_*^2/gD (small scales), due to finite particle path lengths and the delay of aeolian transport adjustment to streamwise shear stress changes.

Because of the lack of flow-field details and those effects on particle mass-transport, the interpretation of drift topography evolution in terms of the flow-field characteristics is difficult. Comparison of drift topography and an oil streak photograph for a cube (Fig. 24 of [11]) show recognizable geometric similarities only for the windward erosional portion. It is difficult to correlate the drift wake topography with the oil-streak wake pattern. Similar comments are appropriate to a comparison of Fig. 2b and Fig. 2a of Baker [13] for the circular cylinder. Oil-streak and smoke [9] photographs show evidence of a so-called horseshoe vortex wrapped around the cylinder and continuing for many diameters downwind. The drift topography photographs, however, show deposit ridges in the near wake which separate the upwind erosional region and the long parallel erosion streaks for the short cylinder and a single erosional streak for the tall cylinder. The surface shear stress must first decrease downwind of the upwind erosional region and then increase again downwind of the ridge. For the very short cylinder, the upwind erosion disappears but the downwind erosional streaks, which have previously been attributed to the "horseshoe vortex", are very much in evidence.

It thus appears that the upwind and downwind erosion are caused by two separate sets of vortex systems. The upwind erosion is obviously caused by the windward portion of the horseshoe vortex. The downwind erosional streaks, at least for the shorter cylinders, are probably caused by separation vortices shed from the leading edges of the upper horizontal surfaces of the cylinders.

Although most previous investigators attribute the persistent twin wake erosional streaks to the horseshoe vortex, Mason and Morton [27] suggest that the dominant trailing vortex pair appears to be a product of the inner wake and is formed by separation of the flow from the lateral and upper surfaces of the obstacle. The horseshoe vortex itself decays rapidly as it is swept around the sides of the element. Although their discussion applies to flow at lower Reynolds number than for the present experiments, the existence of the deposit ridges is in accord with their conclusions regarding details of the flow. The erosional wake streaks are more prominent for the shorter cylinders because the vortices causing the erosion are closer to the surface than for the taller cylinders. The cube does not exhibit the downwind erosional wake streaks because the only upper surface leading edge is normal to the flow and thus a strong windwise upper surface separation vortex is not produced.

It is expected that for a given element geometric shape, that the erosional-depositional pattern is a function of a number of parameters [21], including Froude and Reynolds numbers, density and speed ratios. The primary functional parameter, however, appears to be the aspect ratio, and the data of Figs. 3a, b, 4b clearly show significant geometric and mass rate dependence on the aspect ratio.

The effect of multi-element roughness beds on erodible material emplaced between the elements is to increase the threshold shear stress. This increase in threshold shear stress is a complex function of element shape, aspect ratio and density. For the rectangular blocks circular cylinder and hemisphere arrays represented by the data of Fig. 6, however, the threshold friction speed ratio data are fairly well correlated by the product of drag coefficient and element frontal area per unit floor area.

References

[1] Schlichting, H.: Boundary-layer theory, 6th ed., p. 653, New York: McGraw-Hill (1968).

[2] Wooding, R. A., Bradley, E. F., Marshall, J. K.: Drag due to regular arrays of roughness elements of varying geometry. Boundary-Layer Meteorology 5, 285—308 (1973).

[3] Raupach, M. R., Thom, A. S., Edwards, I.: A wind-tunnel study of turbulent flow close to regularly arrayed rough surfaces. Boundary-Layer Meteorology 18, 373—397 (1980).

[4] Chepil, W. S.: Properties of soil which influence wind erosion. I. The governing principle of surface roughness. Soil Sci. 69, 149—162 (1950).

[5] Lyles, L., Schrandt, R. L., Schmeidler, N. F.: How aerodynamic roughness elements control sand movement. Transactions of the ASAE 17, 134—139 (1974).

[6] Lyles, L., Allison, B. E.: Wind erosion: uniformly spacing nonerodible elements eliminates effects of wind direction variability. J. Soil and Water Conservation 30, 225—226 (1975).

[7] Iversen, J. D.: Small-scale modeling of snow-drift phenomena. In: Wind tunnel modeling for civil engineering applications (T. Reinhold, ed.) Cambridge: University Press, pp. 522—545 (1982).

[8] Roper, A. T.: A cylinder in a turbulent shear layer, Ph. D. Dissertation, Colorado State University, Fort Collins (1967).

[9] Sedney, R.: A survey of the effects of small protuberances on boundary-layer flows. AIAA J. 11, 782—792 (1973).

[10] Hansen, A. C.: Vortex-containing wakes of surface obstacles, Ph. D. Dissertation, Colorado State University, Fort Collins (1975).

[11] Woo, H. G. C., Peterka, J. A., Cermak, J. E.: Wind-tunnel measurements in the wakes of structures. NASA Contractor Report CR-2806, 226 p. (1977).

[12] Hunt, J. C. R., Abell, C. J., Peterka, J. A., Woo, H.: Kinematical studies of the flows around free or surface-mounted obstacles; applying topology to flow visualization. J. Fluid Mech. 86, 179—200 (1978).

[13] Baker, C. J.: The turbulent horseshoe vortex. J. Wind Engineering and Industrial Aerodynamics 6, 9—23 (1980).

[14] Allen, J. R. L.: Scour marks in snow. J. Sedimentary Petrol. 35, 331—338 (1965).

[15] Karcz, I.: Fluviatile obstacle marks from the wadis of the Negev (Southern Israel). J. Sedimentary Petrol. 38, 1000—1012 (1968).

[16] Shen, H. W., Schneider, V. R., Karaki, S.: Local scour around bridge piers. J. Hydraulics Division ASCE 95, 1919—1940 (1969).

[17] Eckman, J. E., Nowell, A. R. M.: Boundary skin friction and sediment transport about an animal-tube mimic. Sedimentology 31, 851—862 (1984).

[18] Iversen, J. D., Greeley, R.: Martian crater dark streak lengths: explanation from wind tunnel experiments. Icarus 58, 358—362 (1984).

[19] Wang, W. P.: Saltation cylinder phenomena: the moire fringe technique. M. S. Thesis, Iowa State University, Ames (1989).

[20] Ostrowski, J. S., Marshall, R. D., Cermak, J. E.: Vortex formation and pressure fluctuations on buildings. Proceedings, International Seminar on Wind Effects on Buildings and Structures, vol. 1. University of Toronto Press, pp. 459—484 (1971).

[21] Iversen, J. D., Wang, W. P., Rasmussen, K. R., Mikkelsen, H. E., Hasiuk, J. F., Leach, R. N.: The effect of a roughness element on local saltation transport. Proceedings of the sixth U.S. National Conference on Wind Engineering (A. Kareem, ed.) A6-1 to A6-10 (1989).

[22] Greeley, R., Iversen, J. D.: Measurements of wind friction speeds over lava surfaces and assessment of sediment transport. Geophys. Res. Lett. 9, 925—928 (1987).

[23] Ng, J. Y.-T.: The structure of the turbulent flow at the test section of BLWT II, ES400, Project Report, University of Western Ontario (1986).

[24] Lettau, H.: Note on aerodynamic roughness-parameter estimation on the basis of roughness-element description. J. Appl. Meteorol. 8, 828—832 (1969).

[25] White, B. R.: Private communication (1989).

[26] Boundy, B., Leach, R.: The effect of surface roughness on flux and threshold velocity of 145 micron sand at atmospheric and Martian pressures. Informal report, NASA, Ames Research Center (1989).

[27] Mason, P. J., Morton, B. R.: Trailing vortices in the wakes of surfacemounted obstacles. J. Fluid Mech. 175, 247—293 (1987).

[28] Gillette, D. A., Stockton, P. H.: The effect of nonerodible particles on wind erosion of erodible surfaces. J. Geophys. Res. 94, 12885—12893 (1989).

[29] Simiu, E., Scanlan, R. H.: Wind effects on structures. New York: John Wiley, p. 137 (1978).

[30] Rae, W. H., Pope, A. Y.: Low-speed wind tunnel testing. New York: John Wiley, p. 167 (1984)

Authors' addresses: J. D. Iversen, Chairman and Professor, and W. P. Wang, Research Assistant, Aerospace Engineering, 304 Town Engineering Building, Iowa State University, Ames, Iowa 50011, USA; K. R. Rasmussen, Geologisk Institut, Aarhus Universitet, Ny Munkegade, Bygn. 520, DK-8000 Aarhus, Denmark; H. E. Mikkelsen, Department of Agrometeorology, Research center, Foulum, DK-8830 Tjele, Denmark, and R. N. Leach, NASA-Ames Research Center, Moffett Field, California 94035, USA.

Acta Mechanica (1991) [Suppl] 2: 77—88

Assessment of aerodynamic roughness
via airborne radar observations

R. Greeley, L. Gaddis, and N. Lancaster, Tempe, Arizona, A. Dobrovolskis, Moffet Field, California, J. Iversen, Ames, Iowa, U.S.A; K. Rasmussen, Aarhus, Denmark; S. Saunders, J. van Zyl, S. Wall, and H. Zebker, Pasadena, California, B. White, Davis, California, U.S.A.

Summary. The objective of this research is to assess the relationship among measurements of roughness parameters derived from radar backscatter, the wind, and topography on various natural surfaces and to understand the underlying physical causes for the relationship. This relationship will form the basis for developing a predictive equation to derive aerodynamic roughness (z_0) from radar backscatter characteristics. Preliminary studies support the existence of such a relationship at the L-band (24 cm wavelength) direct polarization (HH) radar band frequencies. To increase the confidence in the preliminary correlation and to extend the application of the technique to future studies involving regional aeolian dynamics, the preliminary study has been expanded by: 1) defining the empirical relationship between radar backscatter and aerodynamic roughness of bare rocks and soils, 2) investigating the sensitivity of the relationship to microwave parameters using calibrated multiple wavelength, polarization, and incidence angle aircraft radar data, and 3) applying the results to models to gain an understanding of the physical properties which produce the relationship. The approach combines the measurement, analysis, and interpretation of radar data with field investigations of aeolian processes and topographic roughness.

1 Introduction

The ability of wind to initiate particle movement and the flux of windblown sand are dependent on the roughness of the surface (here termed *topographic roughness*), as measured by the equivalent *aerodynamic roughness height* (z_0; described by Bagnold [1], Lyles et al. [2], and others). Typically, z_0 is derived from measurements of the wind velocity profile through the lower atmospheric boundary layer. Consequently, z_0 values for natural surfaces are limited to the site(s) where such measurements are made. Thus, there is a need to determine z_0 values for large areas where aeolian processes occur and that have not been surveyed by traditional means.

This project focuses on the use of aircraft radar data to assess aerodynamic roughness. When microwave radiation is directed at a target surface, the strength of the returned signal depends on the properties of the surface and the characteristics of the radar system. Surface properties affecting the backscatter include: 1) the roughness of the surface at the scale of the radiation's wavelength, 2) the local incidence angle (angle between the surface normal and the incoming radiation) averaged over the spatial extent of the target, and 3) the complex dielectric constant of the target. For surfaces with modest topography at the scale resolvable by the sensor and whose electromagnetic properties are similar, the first effect dominates and the returned signal strength can be taken as a measure of the surface roughness at or near the wavelength scale [3], [4]. The dielectric constant influences the magnitude of the backscatter coefficient and for naturally occurring particulate surfaces is primarily a function of soil moisture content. Important characteristics of the radar system include polarization, wavelength, and incidence angle.

Interactions of microwaves with geologic surfaces consist of coherent and noncoherent surface scattering and volume scattering. Theoretical models employing idealized surface roughness elements are used to relate radar backscatter to surface roughness (e.g., [5]). For all of the models, the size and spacing of the roughness elements responsible for backscatter are dependent on radar wavelength and incidence angle.

Preliminary studies of desert surfaces suggest that a relationship exists between radar backscatter (σ^0) and aerodynamic roughness (z_0) (Fig. 1 and [6]). The relationship between radar backscatter and aerodynamic roughness on land appears to be similar to those governing backscatter from the ocean surface, as measured by Seasat and other oceanographic radar systems (e.g., [7], [8], [9], [10], [11] and others). Although the roughness of the air-sea interface itself depends on wind conditions, the successful recovery of wind and wave patterns at sea from radar data encourages an analogous approach to remote sensing of aerodynamic roughness over land.

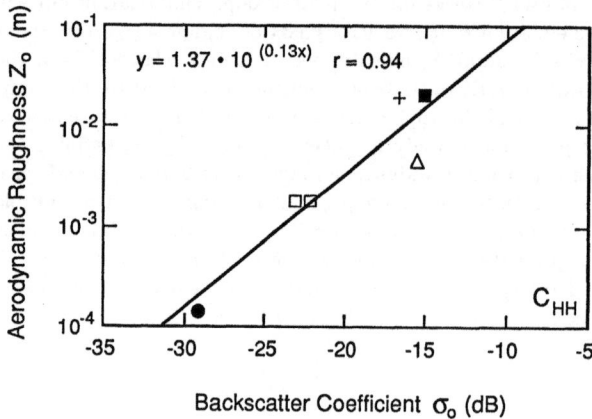

Fig. 1. Relationship between aerodynamic roughness (z_0) and radar backscatter coefficient (σ^0) for 35° incidence angle for C-band; ● playa, □ alluvial fan, + sand-mantled pahoehoe lava, △ pahoehoe lava, ■ aa lava (from [6])

2 Significance and derivation of aerodynamic roughness

In many regions, wind can transport large amounts of sand over long distances, linking zones of deflation and sand supply to depositional areas. Sand accumulations cover 10 to 30% of the surface in many deserts, and constitute important sedimentary systems both today and in the geologic past. Regional or subcontinental aeolian transport systems have been assessed from wind data [12], [13], [14], [15] and confirmed by studies of Landsat and Meteosat images (e.g., [16], [17]). Within these systems, sand is transported over surfaces of different topographic roughness characteristics, including river valleys, sand sheets, dunes, and rock and gravel plains [17], [18], [19], [20].

Estimates of sand transport rates in desert regions [14], [15], [19], [21] are based on equations developed by Bagnold [1], [22] or Lettau and Lettau [23]. These assume a linear relationship between surface shear stress, as expressed by surface friction speed, u_*, and

Table 1. Symbols and abbreviations

C	Mass transport coefficient
D	Zero plane displacement
D_p	Particle diameter
g	Gravitational acceleration
h	Reference height of roughness elements
n	Number of individual roughness elements
q	Mass transport flux of sand
q_0	Mass transport in absence of non-erodible roughness elements
q_r	Mass transport in presence of non-erodible roughness elements
Ri	Richardson number
u_*	Wind friction speed
u_{*t}	Threshold friction speed
u_{*t0}	Threshold friction speed for sand sheet
u_{*tr}	Threshold friction speed in presence of non-erodible roughness elements
U	Wind velocity
z	Height above surface
z_0	Aeolian roughness height
L-, C-, Ku-band	Radar frequencies with wavelengths of ~ 25, $3.8-7.5$, and $1.7-2.4$ cm respectively
HH	Radar polarization: horizontal transmit, horizontal receive polarization
VV	Radar polarization: vertical transmit, vertical receive polarization
HV	Radar polarization: horizontal transmit, vertical receive polarization
VH	Radar polarization: vertical transmit, horizontal receive polarization
θ	Incidence angle (angle between the surface normal and the incoming radar wave)
τ	Surface shear stress
ϱ	Air density
σ^0	Radar backscatter coefficient

wind velocity, U, for a constant value for aerodynamic roughness height, z_0 (symbols defined in Table 1). Use of such assumptions makes it possible to calculate sand transport rates from standard meteorological observations at a single height above the ground. However, such calculated rates may be in error by more than an order of magnitude because of variations in topographic roughness which affect z_0 and u_*.

Aeolian transport of sand results from the drag or shear stress imparted by the wind to the surface. Shear stress causes an increase in wind velocity with height in the turbulent atmospheric boundary layer, and can be written:

$$\tau = \varrho u_*{}^2 \tag{1}$$

where τ is surface shear stress, ϱ is air density and u_* is friction speed, determined from the nature of the wind velocity profile. Following mixing length theory, time averaged wind speed profiles over a natural surface can be expressed by:

$$U = u_*/0.4\big(\ln\big((z - D)/z_0\big)\big) \tag{2}$$

where U is wind velocity, z is a reference height, and D is the zero-plane displacement.

Aerodynamic roughness is a function of the size and spacing of roughness elements on the surface [2], [24], [25]. For closely packed objects, such as sand grains, z_0 is approximately 1/30 their reference height (h). A maximum value of z_0, equal to 1/8 h, is obtained when the roughness elements are spaced at twice their height. As spacing increases, z_0 decreases again. *In situ* measurements of z_0 on geologic surfaces range from < 0.02 cm on playa surfaces to $1-2$ cm on rough lava flows with a local relief of 800 cm [6, 26]. On sand surfaces z_0 is typically on the order of 0.0008—0.0010 cm when no sand transport is taking place, but rises to 0.3—0.8 cm when active sand transport by saltation is occurring [1].

Potential sand transport rates can be calculated from wind data using a variety of formulae which are summarized in Greeley and Iversen [25]. Sand transport rates vary between surfaces of different roughness height because threshold friction speed, u_{*t}, varies with the ratio of the average particle diameter, D_p, to aerodynamic roughness height z_0. It is assumed that the effect of roughness on the saltation threshold is approximately given by:

$$u_{*tr}/u_{*t0} = 2(D_p/z_0)^{-1/5}, \; 1/30 < z_0/D_p < 6 \tag{3}$$

where u_{*t0} is threshold friction-speed in the absence of non-erodible roughness elements and u_{*tr} is the threshold friction-speed in the presence of non-erodible roughness elements. If, for example, the mass transport rate, q, is expressed as:

$$q = C\varrho u_*^2(u_* - u_{*t})/g \tag{4}$$

where ϱ is air density, g is gravitational acceleration, and C is a mass transport coefficient, then the effect of non-erodible roughness on mass transport can be estimated by assuming that the effect is due only to the increase in threshold friction speed. Defining q_r as the mass transport in the presence of non-erodible roughness elements and q_0 as mass transport rate in the absence of non-erodible roughness elements,

$$q_r/q_0 = [u_*/u_{*t0} - 2x(z_0/D_p)^{1/5}]/[u_*/u_{*t0} - 1]. \tag{5}$$

For given values of u_*, the effect of increase in roughness is to decrease the mass transport rate, until a roughness level is reached at which sand transport is no longer sustained (Fig. 2). Equations (3) and (5) should be considered as approximate, as they are based on limited data.

It should be noted, however, that for a given synoptic wind system, an aerodynamically rougher surface exhibits a larger value of surface friction speed, so that mass transport may be larger or smaller on the rougher surface, depending on the relative values of u_*, u_{*t0}, and u_{*tr}. Greeley and Iversen [26] compared sand transport rates on an alluvial fan and lava flow and showed that sand transport occurs only on the smoother alluvial fan surface at low wind velocities; at intermediate wind velocities, transport occurred on both surfaces, with greater sand transport on the alluvial fan; however, at high wind velocities, sand transport rates could be higher on the rough lava surface because of sufficiently higher surface stress due to roughness.

It is clear that aerodynamic roughness (z_0) is an important quantity which strongly influences the absolute and relative rates of sand transport by the wind. Limited data on aerodynamic roughness on different surfaces have been derived from field measurements of wind velocity profiles. Typically, many hours of data recording are necessary to obtain representative wind profiles and, consequently, the existing data set is very small and covers a limited number of surface types. Wind profile measurements are not a part of routine meteorological observations and there is a need to develop a technique for deriving aeolian roughness from remotely sensed data. Radar interacts with roughness elements at the scale of the radar wavelength and may be useful in estimating an aerodynamic roughness parameter. This parameter can then be used to assess aeolian sand transport potential on a regional scale.

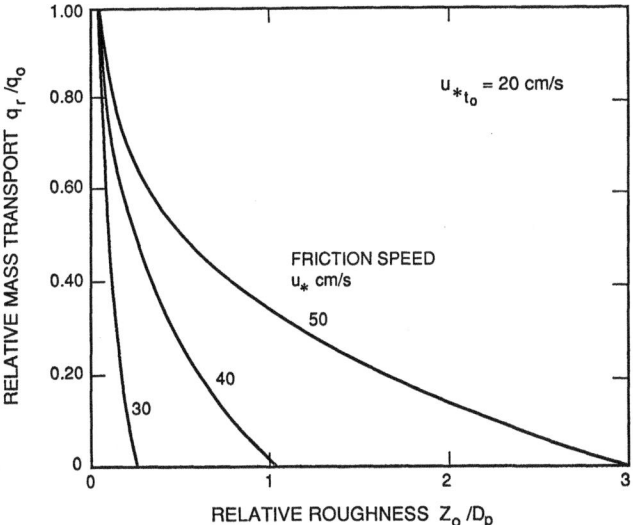

Fig. 2. The influence of surface roughness on the transport of windblown particles. Equation 5 was used to produce this plot (from [5])

3 Surface roughness information from radar backscatter

Understanding the quantitative relationship between radar backscatter coefficient and topographic roughness requires statistical descriptions of the surface and calibrated radar backscatter values [3]. Although the lack of calibrated radar data has hindered efforts to develop a quantitative relationship, some correlations have been reported. For example, statistical measures of topography which are used in conjunction with various theoretical radar models include standard deviation of height (RMS, or root mean square height), correlation length, power density spectra, and RMS slopes (e.g., [3], [5], [27], [28]). Schaber et al. [29] and Evans [30] found a correlation between a single parameter combining different aspects of topographic roughness and relative returned power. Using externally calibrated radar images, Wang et al. [31] also found a strong relationship between RMS height and the scattering coefficient.

Multifrequency and multipolarization radar data increase the ability to discriminate and interpret surface roughness [27], [29], [30], [32], [33], [34], [35], [36], [37]. Because radar interacts with topographic roughness elements at the scale of the radar wavelength, backscatter is strongly dependent on the radar frequency [3], [4], [33], [38], [39]. Using the Rayleigh criterion and a Bragg scattering model the dominant and overall roughness spectrum of a surface can be predicted by comparing and ratioing multiple frequency and polarization data sets [29], [30], [34], [36].

The influence of the incidence angle on radar backscatter for surfaces of different topographic roughness is shown by theoretical backscatter curves in Fig. 3. Reversal in relative magnitude of radar returns occurs between 5° and 30° and affects the ability to discriminate between different topographic roughnesses. Above 30° increased separation between the curves suggests better discrimination of units. However, the decrease in the return signal with increasing angle for slight to medium rough topography suggests that increased discrimination of these surfaces could be gained using several incidence angles.

Our primary radar analysis consists of extracting average backscatter coefficient (σ^0)

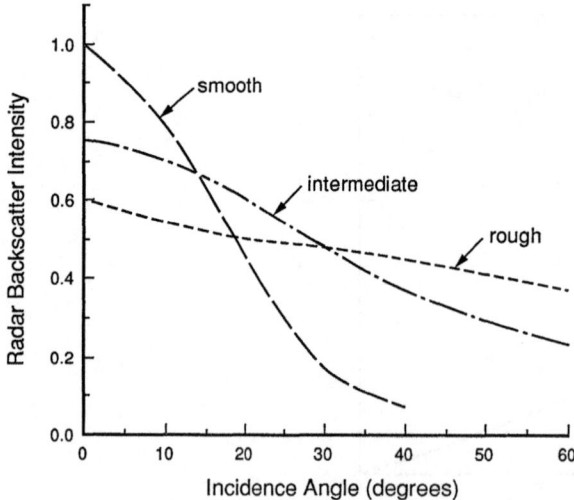

Fig. 3. Generalized radar backscatter curves for surfaces of different roughness at a range of incidence angles (from [5])

values for each surface via digital image processing, following a modification [40] of the technique by Haralick et al. [41] to measure variability of roughness elements across the site observed. Values of σ^0 in the calibrated data are averaged for the site defined on the radar data for all combinations of frequency, polarization, and incidence angle acquired. Ratios of C- and L-band data and like- to cross-polarized data and polarimetry techniques are used to estimate the topographic roughness from backscatter coefficients [34]. These estimates are then compared to statistical measures of topographic roughness. Comparison of the radar data can aid in assessing the ambiguities in radar backscatter as a result of topographic, dielectric constant, and surface penetration effects [34].

From the measured topographic profiles [42], statistical descriptions of the surface are extracted, including root mean square (RMS) heights, slope distributions, correlation lengths, and power spectra. These can be related to the radar scattering characteristics using models currently under development at Jet Propulsion Laboratory [29], [43]. The profiles are also used to quantify the relationship between the distribution of topographic roughness elements and the aerodynamic roughness as given by boundary-layer theory [42]. The results help to clarify the relations among all three observables (radar backscatter, aerodynamic roughness, and topographic relief) and are useful in determining the radar parameters most suitable for predicting aerodynamic roughness.

4 Determination of aerodynamic roughness height, z_0

The goal of the wind measurements is to characterize wind profiles over surfaces of different roughness in order to obtain values for aerodynamic roughness height (z_0), wind friction speed (u_*), and zero-plane displacement (D). The selection of sites for wind profile measurements in this study take into account the need for an adequate fetch of the wind across the surface, so that the boundary layer will be in equilibrium. An upwind distance of approximately 1000 roughness element heights is suggested by Counihan [44] from wind-tunnel studies and of 200 roughness elements by Bradley [45] from field measurements. We took a conservative approach and used sites with a uniform fetch of $\geqq 1000$ roughness elements. Wind profile measurements were made with a 9.6 m high field portable mast, equipped with six to ten cup anemometers. Temperature measurements were made at two heights.

Data were recorded with a microcomputer data-logger system and reduced at Arizona State University. Criteria for determining the validity and quality of the wind profile data include: 1) nearly constant winds for a period of at least 20 minutes from the dominant wind direction as determined from previous observations, analysis of field data, and aerial photographs; and 2) wind velocities in excess of the threshold velocity [> 4 m/sec at the lowest anemometer (0.75 m)].

Wind data were reduced following the methods employed by Greeley and Iversen [26] and Lancaster et al. [42]. The data were corrected for non-neutral atmospheric stability using a Monin-Obukhov similarity model in which the wind profile can be described by: $u_{(z)} = u_*/k[\ln z/z_0 + \psi]$ where $\psi(z/L) = a$ stability function, and L is the Monin-Obukhov stability length [46]; u_* and z_0 were then determined by a least squares fit to the relationship between the corrected height (z') and $u_{(z)}$ using an iterative procedure in which the z_0 for neutral conditions was inserted and successively modified so that the mean value of the constant term in the regression was minimized.

Values of z_0 and u_* for each surface tested were plotted against backscatter coefficients and surface roughness parameters to determine the relationship between the three measurements of rouhgness. The nature of the empirical relationships was determined by regression analysis.

5 Correlation of aerodynamic roughness and radar backscatter

Preliminary study indicates that there is a relationship between aerodynamic roughness and radar backscatter (Fig. 1 and [6]). In the initial study [6], three surface types were assessed, includ ing a smooth playa, an alluvial fan, and a rough lava flow. Backscatter coefficients were obtained from scatterometer data for Pisgah lava field, California. Direct polarized radar data were acquired at L $(\lambda = 19$ cm$)$, C $(\lambda = 6.3$ cm$)$, and Ku $(\lambda = 2.3$ cm$)$ bands by the NASA Johnson Space Center Scatterometer. Lava flows of different textures, some of which are mantled with aeolian sand, were identified from aerial photographs and geologic maps and an average backscatter coefficient was obtained for each. A discriminant analysis technique was used to determine which incidence angles provided the best discrimination between the units. Wind data for a bare playa were also obtained at Lucerne Dry Lake (Sullivan, unpublished data) and for an alluvial fan at a site adjacent to Amboy lava field [26]. Because of the variation in surface roughness which exists on alluvial fans [29], the wind data from Amboy were compared to two alluvial fan surfaces at Pisgah.

Plots of aerodynamic roughness heights (z_0) versus radar backscatter coefficients (σ^0) are shown in Fig. 1 for L- (19 cm wavelength) band at an incidence angle of 35°. The aerodynamic roughness height and backscatter coefficient both increase with the increase in surface roughness from smooth playa to rough lava flow. Plots using 30°, 40°, and 45° incidence angle data yielded similar results. To extend the study, additional radar data of the Pisgah and Amboy volcanic fields, Mojave Desert, California were acquired in June, 1988 by the NASA/JPL Airborne Synthetic Aperture Radar (AIRSAR) during the Mojave Field Experiment (MFE; [47]). These radar data were acquired with a spatial resolution of about 10 m, at 3 wavelengths simultaneously (P-band: 68 cm; L-band: 24 cm; and C-band: 5.6 cm), and at multiple polarizations. Although any combination of transmitted and received polarization states may be synthesized utilizing the complete polarimetric capabilities of the instrument [43], [48], only direct- and cross-polarized data (HH, HV,

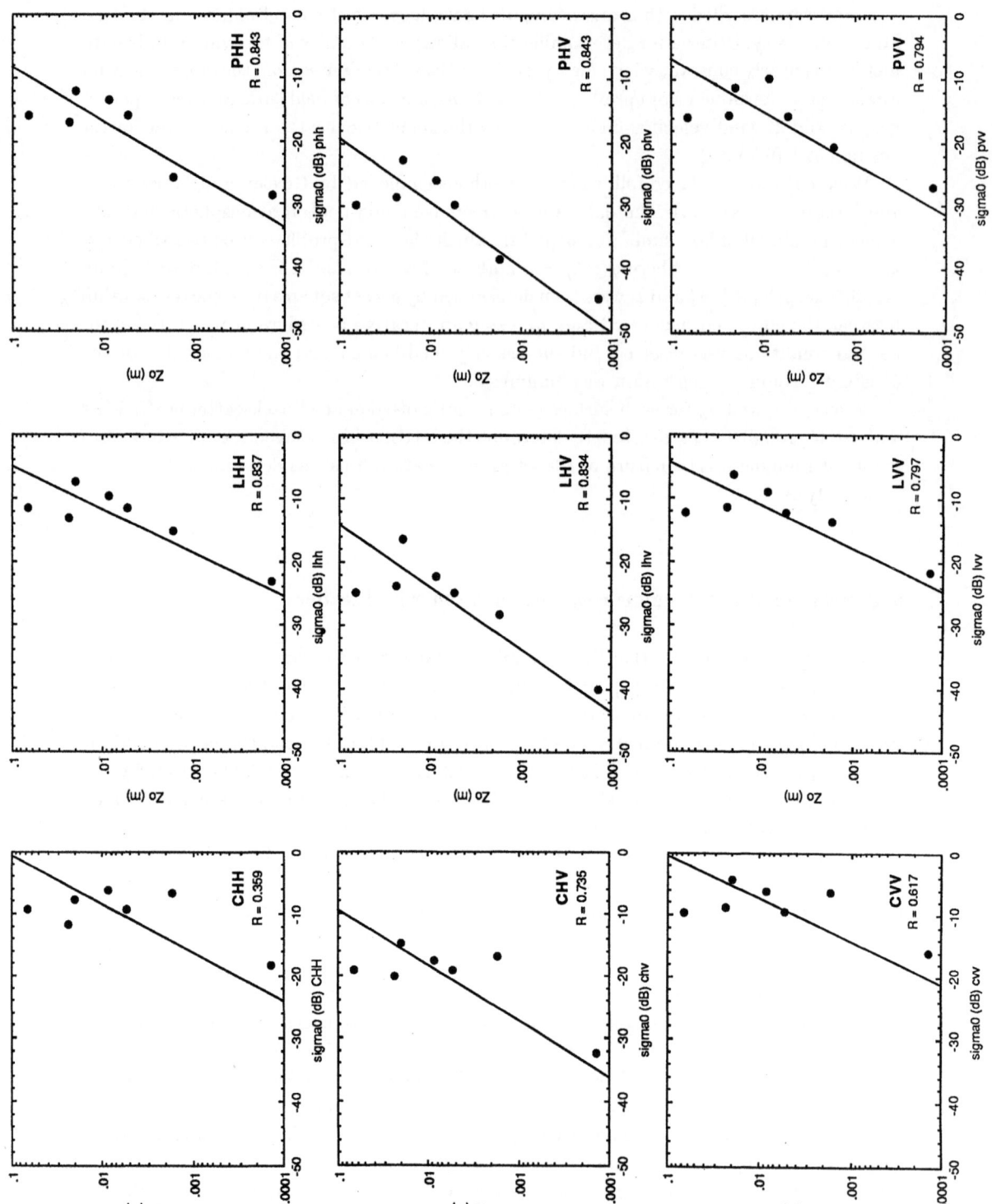

VV, VH) were produced for this analysis. The acquisition of multiple incidence-angle data for a single target was made possible by altering the ground track of the airborne instrument. Images of portions of the Pisgah volcanic field were obtained at center incidence angles of about 30°, 40°, and 50° and of the Amboy field at about 40°.

To assess the correlation between calibrated backscatter coefficients (σ^0) [49] and aerodynamic roughness (z_0), data are shown in Fig. 4 for seven wind tower sites with a range of topographic roughnesses. The general trend of increasing aerodynamic roughness corresponding to increasing backscatter coefficients is observed in all cases. Moderate correlation is observed for the *L*-band and *P*-band backscatter values and the aerodynamic roughness measurements (correlation coefficients, $R \sim 0.80-0.84$). The lowest degree of correlation is observed at *C*-band wavelengths ($R \sim 0.60-0.74$). The relatively high degree of correlation between σ^0 and z_0 at *L*-band and *P*-band wavelengths is attributed to the combined sensitivity of these parameters to topographic roughness on the order of 25 cm, but the influence of subsurface scattering cannot be ruled out at this wavelength. Note that at *C*-band wavelengths the spread of the data points is primarily in the vertical (z_0) direction, indicating that most of the geologic units in this study (except the playa) are approximately equally radar-rough at a scale of ~ 6 cm. At the longer *P*-band wavelength, the spread among the data points is better in both the vertical (z_0) and horizontal (σ^0) directions, but distinction between portions of the backscatter due to surface and subsurface (volume) scattering is difficult. Thus, the longer wavelength radar data may be influenced by surface scatterers ~ 70 cm in size as well as subsurface scatterers up to ~ 70 cm deep, and a correlation between the surface roughnesses reflected in the σ^0 and z_0 may not be valid at this wavelength. Because of these uncertainties, and the small range of correlation coefficients between the direct and cross-polarized backscatter at each wavelength, evaluation of polarization as a factor in these determinations is premature. Additional data, particularly for rougher surfaces, are expected to improve the correlations and to provide greater separation among the radar imaging parameters.

6 Conclusions

Aerodynamic roughness (z_0) is a parameter that is important in the initiation of particle entrainment by the wind and in the flux of windblown particles. This parameter is a function of topographic roughness. The radar backscatter coefficient (σ^0) is also strongly responsive to topography. Consequently, a complex correlation between z_0 and σ^0 could be expected. Results from this and previous studies establish this correlation, with *L*-band radar (wavelength of 24 cm) having the best correlation in comparison to *C*-band (5.6 cm wavelength) and *P*-band (68 cm wavelength). Because *L*-band and *P*-band are most sensitive to topographic roughness on the order of 24 cm, our results suggest that this is the scale of topographic roughness governing or strongly influencing aerodynamic roughness of the arid sites analyzed in this investigation.

Fig. 4. Plots showing the correlation between radar backscatter coefficient (σ^0) versus aerodynamic roughness (z_0) for 7 geologic units (from top to bottom): aa from Pisgah Wind Tower 1 site (incidence angle = 40°); pahoehoe from Pisgah Wind Tower 2 (incidence angle = 40°); mantled pahoehoe from Pisgah Wind Tower 3 (incidence angle = 42°); mantled pahoehoe from Amboy Wind Tower 2 (incidence angle = 34°); alluvium from Amboy Wind Tower 1 (incidence angle = 28°); and playa from Lake Lucerne

This study demonstrates that radar backscatter coefficients obtained from airborne and perhaps orbiting instruments could permit the derivation of aerodynamic roughness values (z_0) for large areas. Such values, when combined with wind frequency data, could enable the assessment of potential aeolian processes on a regional scale. Future work will include additional field studies to collect data for a wide variety of surfaces, measurements of topography, and development of models to explain the correlations among aerodynamic roughness, radar backscatter, and surface topography.

Acknowledgements

This work was supported by the National Aeronautics and Space Administration through the Jet Propulsion Laboratory Contract 88-0407 and Ames Research Center through Grant NCC 2-346.

References

[1] Bagnold, R. A.: The physics of blown sand and desert dunes, 265 pp. London: Chapman and Hall 1941.

[2] Lyles, L., Schrandt, R. L., Schneidler, N. F.: How aerodynamic roughness elements control sand movement. Trans. Am. Soc. Ag. Eng. 17, 134—139 (1974).

[3] Farr, T. G., Engheta, N.: Quantitative comparisons of radar image, scatterometer, and surface roughness data from Pisgah Crater, CA. In: Proc. Int. Geosci. Remote Sensing Symp., San Francisco, CA, 2.1—2.6 (1983).

[4] Blom, R. G., Schenck, L. R., Alley, R. E.: What are the best radar wavelengths, incidence angles, and polarizations for discrimination among lava flows and sedimentary rocks? A statistical approach. IEEE Trans. Geosci. Rem. Sens. GE-25, 208—213 (1987).

[5] Ulaby, F. T., Moore, R. K., Fung, A. K.: Microwave remote sensing, active and passive, vol. 2. Radar remote sensing and surface scattering and emission theory. Reading, Massachusetts: Addison-Wesley Publishing Co. 1982.

[6] Greeley, R., Lancaster, N., Sullivan, R. J., Saunders, R. S., Theilig, E., Wall, S., Dobrovolskis, A., White, B. R., Iversen, J. D.: A relationship between radar backscatter and aerodynamic roughness: preliminary results. Geophys. Res. Letts. 5, 565—568 (1988).

[7] Krishen, K.: Correlation of radar backscattering cross sections with ocean wave height and wind velocity. J. Geophys. Res. 76, 6528—6539 (1971).

[8] Jones, W. L., Schroeder, L. C.: Radar backscatter from the ocean: dependence on surface friction velocity. Boundary Layer Meteorol. 13, 133—149 (1978).

[9] Moore, R. K., Fung, A. K.: Radar determination of winds at sea. Proc. IEEE 67, 1504—1521 (1979).

[10] Jones, W. L., Boggs, D. H., Bracalente, E. M., Brown, R. A., Guymer, T. H., Shelton, D., Schroeder, L. C.: Evaluation of the Seasat wind scatterometer. Nature 294, 704—707 (1981).

[11] Liu, W. T., Large, W. G.: Determination of surface stress by Seasat-SASS: a case study with JASIN data. J. Phys. Oceanogr. 11, 1603—1611 (1981).

[12] Dubief, J.: Le vent et la deplacement du sable au Sahara. Travaux, Institute de Recherches Sahariennes 8, 123—162 (1952).

[13] Brookfield, M.: Dune trend and wind regime in central Australia. Geomorphologie [Suppl. 10]: 121—158 (1970).

[14] Wilson, I. E.: Desert sandflow basins and a model for the development of ergs. Geogr. J. 137, 180—197 (1971).

[15] Lancaster, N.: Winds and sand movements in the Namib sand sea. Earth Surface Processes and Landforms 10, 607—619 (1985).

[16] El Baz, F., Wolfe, R. W.: Wind patterns in the Western Desert. In: El Baz, F. et al. (eds.) Desert landforms of southwestern Egypt: a basis for comparison with Mars. National Aeronautics and Space Administration, Contractor Rep. 3611, 119—140 (1982).

[17] Mainguet, M.: Space observations of Saharan aeolian dynamics. In: El Baz, F. (ed.), Deserts and arid lands. The Hague: Martinus Nyhoff Publ., 31—58 (1984).

[18] Breed, C. S., Fryberger, S. C., Andrews, S., McCauley, C., Lennartz, F., Geber, D., Horstmann, K.: Regional studies of sand seas using LANDSAT (ERTS) imagery. In: McKee, E. D. (ed.) A study of global sand seas. U. S. Geol. Surv. Prof. Paper 1052, 305—398 (1979).

[19] Fryberger, S. G., Ahlbrandt, T. S.: Mechanisms for the formation of eolian sand seas. Z. Geomorph. 23, 440—460 (1979).

[20] Mainguet, M.: The influence of trade winds, local air masses and topographic obstacles on the aeolian movement of sand particles and the origin and distribution of ergs in the Sahara and Australia. Geoforum 9, 17—28 (1978).

[21] Finkel, H. J.: The barchans of southern Peru. J. Geol. 67, 614—647 (1959).

[22] Bagnold, R. A.: Forme des dunes de sable et regime des vents. Actions Eoliennes, Centre National de Recherches Scientifiques, Colloques Internationaux 35, 23 -32 (1953).

[23] Lettau, H., Lettau, K.: Experimental and micrometeorological studies of dune migration. In: Lettau, H., Lettau, K. (eds.) Exploring the world's driest climate. University of Wisconsin-Madison, Institute for Environmental Studies Report 101, 101—147 (1978).

[24] Iversen, J. D., Greeley, R., Pollack, J. B., White, B. R.: Simulation of martian eolian phenomena in the atmospheric wind tunnel. Space Simulation. NASA Special Publication 36, 191—213 (1973).

[25] Greeley, R., Iversen, J. D.: Wind as a geological process on Earth, Mars, Venus and Titan, 333 pp. Cambridge: Cambridge Univ. Press 1985.

[26] Greeley, R., Iversen, J. D.: Measurements of wind friction speeds over lava surfaces and assessment of sediment transport. Geophys. Res. Lett. 14, 925—928 (1987).

[27] Schaber, G. G., Elachi, C., Farr, T. G.: Remote sensing data of SP Mountain and SP lava flow in north-central Arizona. Rem. Sens. of Envir. 9, 149—170 (1980).

[28] Schaber, G. G., Berlin, G. L., Pike, R. J.: Terrain analysis procedures for modeling radar backscatter. In: Radar geology: an assessment. Jet Prop. Lab. Publ. 80—61, 168—181 (1980).

[29] Schaber, G. G., Berlin, G. L., Brown, W. E., Jr.: Variations in surface roughness within Death Valley, California: geological evaluation of 25-cm-wavelength radar images. Geol. Soc. Am. Bull. 87, 29—41 (1976).

[30] Evans, D. L.: Radar observations of a volcanic terrain: Askja Caldera, Iceland. Jet Propulsion Lab. Publ. 78- 81, 39 pp. (1978).

[31] Wang, J. R., Englmann, E. T., Shiue, J. C., Rusek, M., Steinmeier, C.: The SIR-B observations of microwave backscatter dependence on soil moisture, surface roughness, and vegetation covers. IEEE Trans. Geosci. and Rem. Sens. GE-24, 510—516 (1986).

[32] Dellwig, L. F., Moore, R. K.: The geological value of simultaneously produced like- and cross-polarized radar imagery. J. Geophys. Res. 71, 3597—3601 (1966).

[33] Dellwig, L. F.: An evaluation of multifrequency radar imagery of the Pisgah Crater area, California. Mod. Geol. 1, 65—73 (1969).

[34] Daily, M., Elachi, C., Farr, T., Stromberg, W., Williams, S., Schaber, G.: Application of multispectral radar and Landsat imagery to geologic mapping in Death Valley. Jet Prop. Lab. Publ. 78—19, 47 pp. (1978).

[35] Daily, M., Elachi, C., Farr, T., Schaber, G.: Discrimination of geologic units in Death Valley using dual frequency and polarization imaging radar data. Geophys. Res. Lett. 5, 889—892 (1978).

[36] Malin, M. C., Evans, D., Elachi, C.: Imaging radar observations of Askja Caldera, Iceland. Geophys. Res. Lett. 5, 931—934 (1978).

[37] Evans, D. L., Farr, T. G., Ford, J. P., Thompson, T. W., Werner, C. L.: Multipolarization radar images for geologic mapping and vegetation discrimination. IEEE Trans. Geosci. and Rem. Sens. GE-24, 246—257 (1986).

[38] Elachi, C., Blom, R., Daily, M., Farr, T., Saunders, R. S.: Radar imaging of volcanic fields and sand dune fields: implications for VOIR. In: Radar geology: an assessment. Jet Prop. Lab. Publ. 80—61, 114—150 (1980).

[39] Blom, R. G.: Effects of variation in incidence angle and wavelength in radar images of volcanic and aeolian terranes. Int. J. Rem. Sens. (to be submitted).

[40] Gaddis, L., Mouginis-Mark, P., Singer, R., Kaupp, V.: Geologic analyses of Shuttle Imaging Radar (SIR-B) data of Kilauea Volcano, Hawaii. Bull. Geol. Soc. Am. 101, 317—332 (1989).

[41] Haralick, R. M., Shanmugam, K., Dinstein, I.: Textural features of image classification. IEEE Trans, Systems, Man, and Cybernetics SMC-3, 610—621 (1973).

[42] Lancaster, N., Greeley, R., Rasmussen, K.: Interaction between unvegetated desert surfaces and the atmospheric boundary layer: a preliminary assessment (this issue), (1990).

[43] van Zyl, J. J., Zebker, H. A., Elachi, C.: Imaging radar polarization signatures: theory and observations. Radio Sci. 22, 529—543 (1987).

[44] Counihan, J.: Wind tunnel determination of the roughness length as a function of three dimensional roughness elements. Atmos. Envir. 5, 637—642 (1971).

[45] Bradley, E. F.: A micrometeorological study of velocity profiles and surface drag in the region modified by a change in surface roughness. Quart. J. Royal Meteor. Soc. 94, 361—379 (1968).

[46] Fleagle, R. G., Businger, J. A.: An introduction to atmospheric physics. 432 pp. New York: Academic Press 1980.

[47] Wall, S., van Zyl, J. J., Arvidson, R. E., Theilig, E., Saunders, R. S.: The Mojave field experiment: precursor to the planetary test site (abstract). Bull. Am. Astronom. Soc. 20, 809 (1988).

[48] Zebker, H. A., van Zyl, J. J., Held, D. N.: Imaging radar polarimetery from wave synthesis. J. Geophys. Res. 92, 638—2701 (1987).

[49] van Zyl, J. J.: Calibration of polarimetric radar images using only image parameters and trihedral corner reflector responses. IEEE Trans. Geosci. Remote Sens. 28, 337—348 (1990).

Authors' addresses: R. Greeley, L. R. Gaddis, N. Lancaster, Department of Geology, Arizona State University, Tempe, AZ 85287-1404;

A. Dobrovolskis, National Aeronautics and Space Administration, Ames Research Center, Mail Stop 245-3, Moffett Field, CA 94035-1000;

J. D. Iversen, Department of Aerospace Engineering, Iowa State University, Ames, IA, 50010, U.S.A.;

K. Rasmussen, Institute of Geology, Aarhus University, 520 Ny Munkegade, DK-8000 Aarhus C, Denmark;

R. S. Saunders, J. J. van Zyl, S. D. Wall, H. A. Zebker, Jet Propulsion Laboratory, 4800 Oak Grove Drive, Pasadena, CA 91109;

B. R. White, Department of Mechanical Engineering, University of California at Davis, Davis, CA 95616, U.S.A.

Acta Mechanica (1991) [Suppl] 2: 89—102
© by Springer-Verlag 1991

Interaction between unvegetated desert surfaces and the atmospheric boundary layer: a preliminary assessment

N. Lancaster, R. Greeley, and **K. R. Rasmussen***, Tempe, Arizona

Summary. The nature of interactions between surface winds and natural desert surfaces has important implications for aeolian sediment transport. We report here initial results from measurements of boundary layer wind profiles and surface roughness at 5 sites in Death Valley, U.S.A. and discuss their implications. The sites studied were a flat to gently undulating gravel and sand reg, three alluvial fan surfaces, including one with a well-developed desert pavement, and a silt and clay playa. Aerodynamic roughness estimates range from 0.000 18 to 0.005 37 m and increase in parallel with the visual estimates of topographic roughness at each site. Microtopography was measured with template and laser profiling devices. The standard deviation of surface elevations (RMS height) appears to provide a good index of surface roughness. It correlates well with field observations of topographic roughness and aerodynamic roughness estimates.

1 Introduction

The nature of interactions between surface winds and natural desert surfaces has important implications for understanding sediment mobilization and transport by the wind [1], [2]. The topographic and particle size roughness of the surface determines its aerodynamic roughness length (z_0) [3]. This in turn affects friction speed (u_*) and therefore surface shear stress (τ) and the flux of sediment that can be mobilized from or transported across the surface [1], [4].

Despite the importance of a good understanding of the effects of surface roughness on aeolian sediment transport to fields such as desertification, regional-scale sediment flux, and dune initiation, there have been few detailed field studies of the effects of surface roughness on boundary layer development and aeolian sediment transport over natural desert surfaces, especially those that are composed of a high percentage of immobile or nonerodible particles (e.g. [4], [5]). We report here initial results from measurements of boundary layer wind profiles and surface roughness at 5 sites in Death Valley, U.S.A. The investigations were conducted as part of an ongoing series of field, wind tunnel, and theoretical studies of the relationship between aerodynamic roughness and radar backscatter of surfaces over which aeolian sediment transport takes place as part of the NASA Shuttle Imaging Radar (SIR-C) Mission. The field studies reported here were designed to assess the range of aerodynamic roughness (z_0) over natural erodible surfaces and to determine and quantify properties of the surface that may influence aerodynamic roughness.

---

* On sabbatical from Institute of Geology, Aarhus University, DK-8000 Aarhus, Denmark.

2 The study sites

Studies were carried out at the following sites in Death Valley National Monument:
Stovepipe Wells, Kit Fox Fan, Trail Canyon Fan, Golden Canyon Fan, and Confidence
Mill Playa (Fig. 1). The sites were selected to span a range of aeolian and radar roughness
values, known or potential sources for windblown sand or dust, and surfaces over which
sand and dust are transported. Each site was located to minimize or avoid the effects of
large scale topography on the wind, and to maximize fetch in the direction of prevailing
sand and dust transporting winds. The sites chosen were free of vegetation and as uniform
in terms of surface characteristics as possible.

Site characteristics

The Stovepipe Wells site (Fig. 2A) is on a flat to gently undulating sand and gravel surface (reg) on the distal part of an alluvial fan. Small (10—20 cm deep, 1—2 m wide) washes
cross the surface, which at the time of investigations was completely unvegetated. Average
slope is $< 1°$ to the south, and 1—2° from east to west. Active transport of sand across this
surface is evidenced by 10—20 cm high sand drifts with wind ripples in the washes. In
addition, active transport of both sand and dust was observed. Some of the clasts on the
surface show evidence of wind facetting.

Kit Fox Fan site (Fig. 2B) is located on the active mid- to proximal part of the
same fan as the Stovepipe Wells site. Its surface consists of gravel to cobble sized clasts,
with a sandy matrix. Average slope of the site is 0.5° to the southeast, and 3° to the west.
Multiple small channels (5—10 cm deep, 1—2 m wide) floored with sand and granules

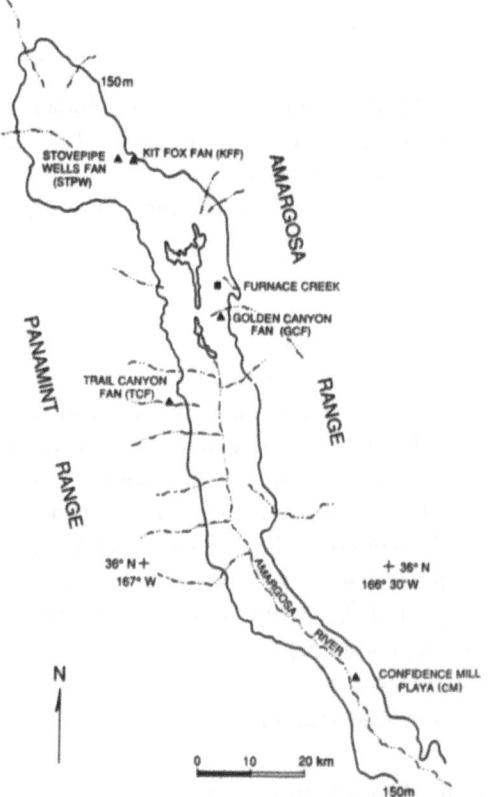

Fig. 1. Sketch map to show the location of the
study sites in Death Valley National Monument

Fig. 2. Photographs of the study sites: **A** Stovepipe Wells: view southwest toward edge of Stovepipe Wells dunefield. Note thin sand cover and small shadow dune in wash in center of view. **B** Kit Fox Fan: view west (down fan) to Stovepipe Wells dunefield. Note rough surface of fan, and anastomosing washes. **C** Golden Canyon Fan: view southeast across fan. Note large cobbles and boulders on bars, with coarse sand and gravel in swales. **D** Trail Canyon Fan: view north across fan. Note well-developed desert pavement surface. **E** Confidence Mill playa: view north toward main part of Death Valley. Note puffy surface of playa

cross the site from east to west. Scattered ephemeral herbaceous plants > 1 m apart were observed at this site. Active sand transport was observed in the field.

The Golden Canyon Fan site (Fig. 2 C) is on an active alluvial fan/mudflow surface and has a well developed bar and swale topography of 20—50 cm relief. Channels (swales) are generally covered by sandy or silty material. Bars are composed of gravel to small boulders of a variety of lithologies set in a sandy silt matrix. Average slopes are $< 1°$ to the southeast, and $2°$ downfan (west). There was no vegetation at this site.

The site at Trail Canyon Fan (Fig. 2D) is on a well developed desert pavement on the lower part of the alluvial fan. The surface consists of interlocking oblate gravel to cobble-sized clasts, with scattered boulders. A well-developed desert varnish coats all of the clasts. The surface is gently undulating to flat and is incised by 1.5 m deep washes. Spacing of washes is more than 100—200 m. Bar and swale topography is evident in places. Average slopes are $< 1°$ to the north, and $3.5°$ to the east (down fan). There is no vegetation except along the washes where scattered creosote bushes occur.

Confidence Mill Playa site (Fig. 2E) is on a clay-silt playa (dry lake bed) that forms the terminus of the Amargosa River. The surface is flat with a soft to locally hard puffy clay-rich crust, and scattered south-north trending "channels" 5—10 cm deep.

3 Boundary layer wind profile studies

Measurements of surface winds and boundary layer profiles were carried out at each site for a period of approximately 28 days during April and May 1989.

Instrumentation

Boundary layer wind profiles were measured using field-portable anemometer masts with a height of 9.8 m. Cup anemometers were placed at the following heights with a logarithmic spacing: 0.75, 1.25, 2.07, 3.44, 5.72, and 9.5 m. Pairs of temperature sensors were placed in a shielded and ventilated mounting at heights of 1.3 and 9.6 m. Wind directions were measured with wind vanes at heights of 9.7 m and 1.5 m. At sites that were considered to have the largest topographic roughness (Golden Canyon Fan, Kit Fox Fan), an additional mast with four anomemeters at heights of 1.0, 2.11, 4.49 and 9.5 m was sited 80 m from the main mast. Data were recorded using a data logger with a sampling interval of 25 seconds, and averaged for a 20 minute period.

Wind data reduction and analysis

Data were initially sorted by the wind speed at the anemometer closest to the ground (0.75 m). A subset of the data was extracted in which wind speeds at all anemometers were above 4 m/sec. More than 200 wind profiles with all wind speeds > 4 m/sec from three major directional sectors (N—NE, SE—S, W—NW) were obtained at each site (see Fig. 3 for examples of profiles obtained in neutral conditions). The > 4 m/sec data were sorted by directional sector and stability characteristics after calculation of the bulk Richardson number (Ri).

Wind profiles in a horizontally homogenous and thermally neutral surface can be described by:

$$u(z) = u_*/\varkappa \ln z - d_0/z_0 \tag{1}$$

where $\varkappa$ is von Karmann's constant, z_0 is the aerodynamic roughness length and d_0 is the zero plane displacement. Since the surfaces studied in Death Valley are without vegetation and have small or moderately sized roughness elements, d_0 can be ignored.

However, most of the data were obtained in conditions of strong incoming solar radia-

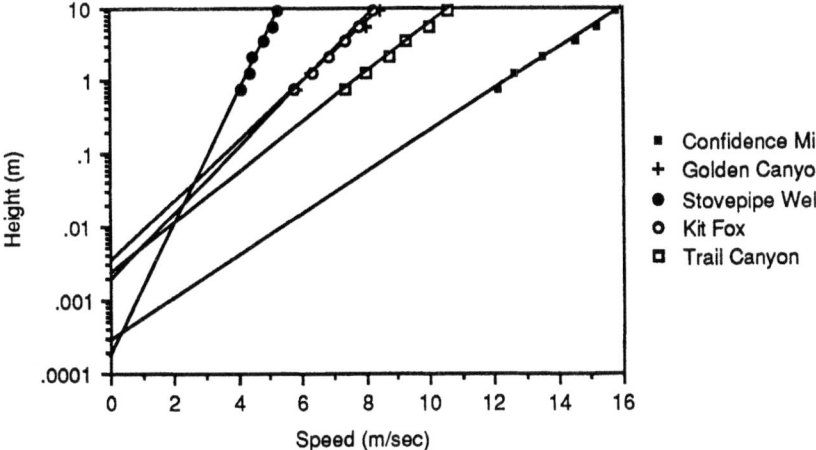

Fig. 3. Sample wind profiles (neutral conditions) for each of the study sites

tion. Therefore buoyancy influences on the wind profile must be considered. The stability correction can be expressed in terms of the buoyancy length scale $\zeta = z/L$ where L is the Monin-Obukhov length [6] so that the wind profile in non-neutral conditions can be expressed as:

$$u(z) = u_*/\varkappa[\ln z/z_0 + \psi] \tag{2}$$

where, following [7], [8], the stability function $\psi(z/L)$ is:

$$\psi(z/L) = 1.1(-z/L)^{0.5} \qquad \text{(unstable)} \tag{3}$$

and

$$\psi(z/L) = -4.8(-z/L) \qquad \text{(stable)}. \tag{4}$$

We did not measure the Monin-Obukhov length scale directly, but derived it from a relationship to the simpler Richardson number (Ri):

$$\text{Ri} = g(\partial T/\partial z)/T_0(\partial u/\partial z)^2 \tag{5}$$

where g = gravitational acceleration (9.81 m/sec/sec); $\partial T/\partial z$ = the average vertical gradient of potential temperature; and $\partial u/\partial z$ is the vertical gradient of the mean wind speed. Temperatures were only measured at two heights in the field (1.3 and 9.6 m), so the Richardson number can be approximated by:

$$\text{Ri} = g\varDelta Tz/T_0\varDelta u^2 \tag{6}$$

where $\varDelta T$ is the temperature difference between the lower and upper temperature sensors, expressed as a potential temperature; z = the geometric mean height of the temperature sensors; T_0 = the surface reference temperature in °K (approximated by the lower temperature sensor reading); and $\varDelta u$ is the difference in wind speeds between the anemometers at 0.75 and 9.6 m [9]. Corrections were made so that:

$$z' = \ln z_i - \psi \qquad \text{(unstable)} \tag{7}$$

and

$$z' = \ln z_i + \psi \qquad \text{(stable)}. \tag{8}$$

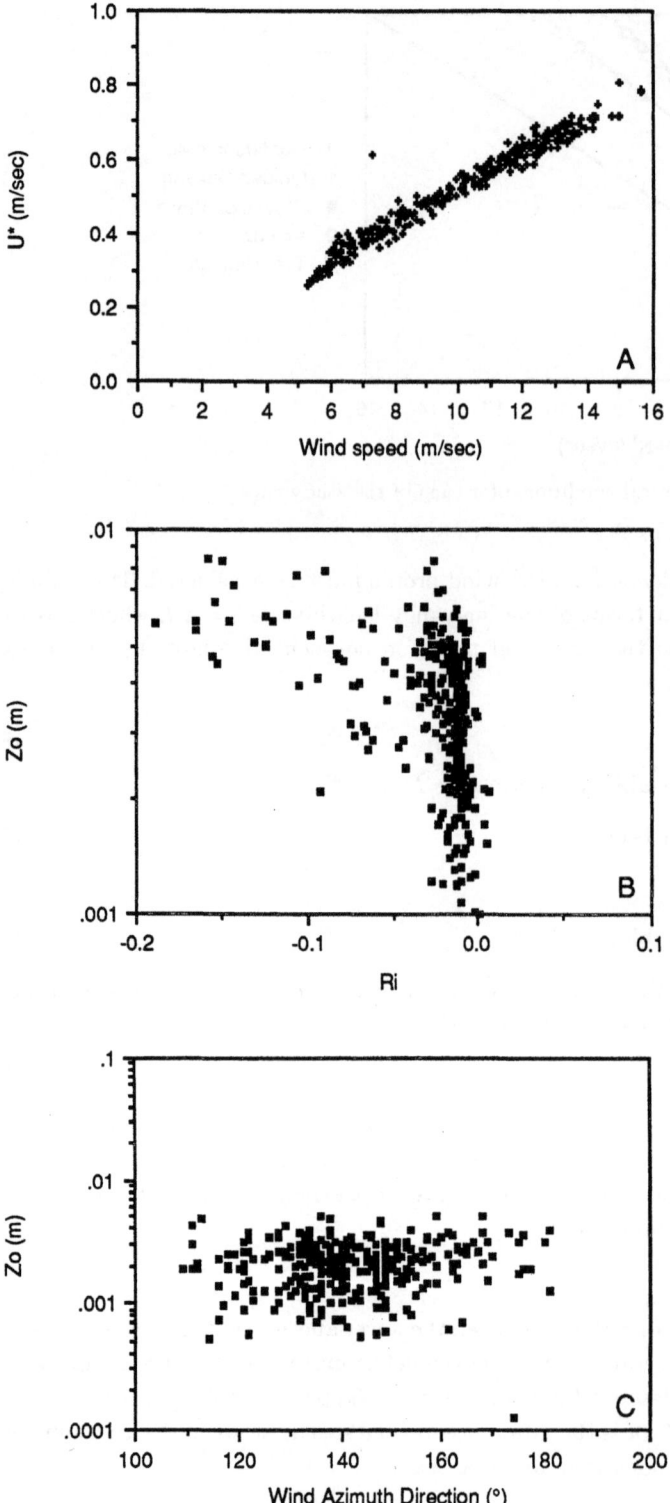

Fig. 4. Relationships amongst wind profile parameters. Examples selected from southerly wind data at Golden Canyon Fan. **A** Relationship between friction speed (U^*) and wind speed at 9.6 m. **B** Relationship between aerodynamic roughness (z_0) and bulk Richardson number (Ri). **C** Relationship between aerodynamic roughness (z_0) and wind azimuth direction (°)

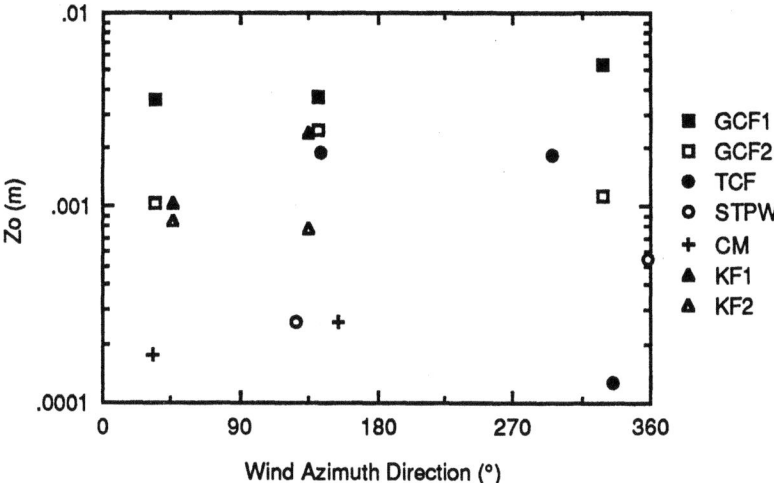

Fig. 5. Comparison between mean values of estimates of aerodynamic roughness (z_0) for sites studied in Death Valley. *GCF* Golden Canyon Fan, *TCF* Trail Canyon Fan, *STPW* Stovepipe Wells, *CM* Confidence Mill, *KF* Kit Fox Fan

$U*$ and z_0 were then determined by a least squares fit to the relationship between the corrected height (z') and $u_{(z)}$ using an interative procedure in which a value of z_0 at neutral stratification was inserted and successively modified so that the mean value of the constant term in the regression was minimized.

Several checks were placed on the data to ensure the reliability and consistency of the estimates of wind profile parameters. Plots of $u*$ versus wind speed at 9.5 m height (Fig. 4A) show that in all cases friction speed increases with wind speed. Plots of z_0 versus Richardson number (Fig. 4B) show that there was a slight tendency for z_0 to increase with increasing instability. One possible explanation may be imperfect shielding of the temperature probes by our naturally ventilated screens, giving rise to eroneous estimates for Ri. Plots of z_0 versus wind direction (e.g. Fig. 4C) indicate that, in most cases, z_0 varies within 10—20% for each directional sector.

Estimates of aerodynamic roughness and friction speed

Table 1 gives the mean values aerodynamic roughness obtained for each major directional sector at each site. Comparisons between the values of aerodynamic roughness obtained for each site are shown in Fig. 5. They indicate that aerodynamic roughness estimates increase from the smoothest site, Confidence Mill playa, to the roughest site, Golden Canyon Fan, in parallel with the visual estimates of topographic roughness at each site.

There are some differences in the estimates of z_0 for different wind directions, suggesting that local site conditions may strongly affect aerodynamic roughness estimates. The largest difference is at Trail Canyon, where the z_0 estimate for northerly winds is an order of magnitude less than that for southerly and westerly directions. At sites where two masts were used, there are also some differences between z_0 values for each mast. At Golden Canyon estimates of z_0 for the mast in the more distal location are consistently lower than those for the main mast upfan, suggesting perhaps that the site was not as homogenous as was thought. At Kit Fox, values for the two masts are almost identical for northerly winds, but diverge significantly for southerly winds.

Table 1. Arithmetic mean values of aerodynamic roughness (z_0). Units in meters

Site	Wind direction			
	N—NE	SE-S	W—NW	NW—NNW
Stovepipe Wells		0.00026	0.00055	
Kit Fox				
Mast 1	0.00107	0.00237		
Mast 2	0.00085	0.00078		
Golden Canyon				
Mast 1	0.00356	0.00360	0.00537	
Mast 2	0.00110	0.00245	0.00113	
Trail Canyon		0.00190	0.00182	0.00012
Confidence Mill	0.00063	0.00018		

4 Surface roughness studies

The surface roughness of natural unvegetated surfaces can be regarded as consisting of two components: (1) the size and arrangement of surface particles, and (2) the microtopography of the surface.

Surface particle size

The sizes of surface particles was assessed by estimating the proportions of clay/silt, sand, gravel, cobbles and boulders in a 1 m square area at 4 or 5 locations chosen to represent the range of materials at each site within a 24×24 m square centered on the anemometer mast. The average proportions (% of surface covered) of different sized materials at each site are given in Table 2. Kit Fox Fan and Stovepipe Wells are dominated by sand- and gravel-sized particles, Trail Canyon Fan is composed of gravel and cobbles and is the "best sorted", with the most uniform particle sizes. This reflects its well developed desert pavement and relatively greater age and geomorphic stability. The most variable site was Golden Canyon, which is on a recently active alluvial fan/debris flow.

Microtopography

In this study we defined microtopography as centimeter to decimeter scale surface relief. Microtopographic measurements were made using two devices, both of which provided data from which a detailed profile of the ground surface could be reconstructed: a template, and a laser-photo device. The template (Fig. 6) consisted of a 2 m long horizontal bar through which 200 1 m-long by 0.5 cm-diameter vertical aluminium rods protruded,

Table 2. Estimates of average particle size composition of Death Valley Sites

Site	Mean particle size composition (% of surface covered)				
	Clay/silt	Sand	Gravel	Cobbles	Boulders
Stovepipe Wells	1.3	42.5	45.0	11.5	
Kit Fox Fan		63.0	20.0	11.0	6.0
Golden Canyon Fan	25.0	23.0	28.3	17.0	6.7
Trail Canyon Fan			65.0	20.0	15.0
Confidence Mill Playa	100.0				

giving a horizontal resolution of 1 cm. These rods were allowed to slide to contact the ground surface when the bar adjusted to a horizontal position with a carpenter's bubble level. The positions of the rods were recorded photographically. The height of the bar above the ground surface was also measured at each end, so that adjacent template measurements could be linked together. In this manner two profiles 24 m long were obtained approximately parallel to and perpendicular to the major wind directions. In the laboratory, each photograph was printed to the same scale, and the position of the tops of each rod were digitized with an accuracy of ~ 2 mm. However, the rods may in some cases protrude into soft spots on the bed, and we consider that the overall accuracy is about 5 mm. The template method was employed at all the rougher sites. At sites where the ground surface was composed of soft silt or sand (Stovepipe Wells and Confidence Mill) it was found that the rods penetrated the ground surface to an unacceptable degree and gave eroneous measurements.

The laser profiling device (Fig. 7) consisted of a triangle with sides 1.2 m long and raised above the ground by about 50 cm. A small laser was mounted vertically on a traversing stage that was moved by a small motor along the rails that formed each side of the device. The traverse time was set to approximately 45 seconds. To operate the device and produce a record of microtopography, a camera was mounted on the vertex opposite the laser stage, and a time exposure of ~ 45 seconds made as the laser traversed the opposite side. A small electronic flash was used to illuminate a scale and the ground surface. This produced an image of the ground surface with a red profile line imaged across it by the laser. The camera angle was ~ 30° to the horizontal. This procedure was repeated for each side of the triangle, and the images were then digitized to produce a topographic profile with a 2—3 mm horizontal and vertical resolution.

The digital terrain data produced by both techniques were then analyzed to produce terrain statistics after the linear trend in the data due to overall surface slope was removed. The RMS height (standard deviation of the changes in elevation) and correlation length were then calculated from the unfiltered data, for which the longest possible wavelength (1) will be half the transect distance (12 m). The results are given in Table 3. Comparison of the data shows that the RMS height of the laser data is about an order of magnitude less than that of the template data, suggesting that it is measuring particle roughness.

These data show that roughness, as measured by RMS height derived from both systems,

Fig. 6. Template in use at Stovepipe Wells. Data were digitized from a photograph similar to this, scaled using the scale bar in the center of the photograph

Fig. 7. Laser photographic profile device. Note traversing stage with laser on left and camera in center

increases in the same sense as that suggested by visual inspection of the surface. Both the template and particle size data suggest that Golden Canyon Fan was the roughest site studied. Kit Fox Fan and Trail Canyon Fan differ only slightly from each other. The E—W roughness is less in all cases than the N—S values. This is probably a result of the E—W orientation of the bar and swale topography and small washes that occur at all sites. As the winds cross this topography at an oblique angle, a further index of the surface roughness is the geometric mean of the N—S and E—W RMS heights (Table 3).

Table 3. RMS height (m) derived from template and laser profiles

	N—S	E—W	Geometric mean
Stovepipe Wells			
Laser			0.0056
Kit Fox Fan			
Template	0.0615	0.0329	0.0420
Laser			0.0097
Golden Canyon Fan			
Template	0.0772	0.0394	0.0571
Laser			0.0150
Trail Canyon Fan			
Template	0.0660	0.0256	0.0364
Laser			0.0076
Confidence Mill			
Laser			0.0066

5 Comparisons between estimates of aerodynamic roughness and surface roughness measurements

The aerodynamic roughness of a surface is a function of its surface roughness characteristics. These can be sub-divided into two components: particle roughness, which is a function of surface particle size and spacing, and topographic roughness, which is a function of the microtopography of the site. On a horizontal surface with uniform bed material, aerodynamic roughness can be approximated by $d/30$ where d is the mean particle diameter [10]. Figure 8 shows that there is a poor correlation ($r = 0.48$) between the mean particle size at each site and the aerodynamic roughness estimates. This suggests that, for surfaces with a mix of particle sizes, the mean particle size is not a good measure of surface roughness. Factors such as particle spacing are probably important. In addition, the particle roughness is superimposed on the microtopography of the surface.

The template and laser data can be analysed to provide a measure of roughness at both scales. The RMS height of the surface appears to be a good index of its overall roughness. The RMS height was compared to the aerodynamic roughness estimates for both northerly and southerly wind directions at each site. The plots of these data (Fig. 9) show that there is a good relationship ($r = 0.97$, N—S; $r = 0.63$, E—W) between unfiltered RMS height derived from the template data and aerodynamic roughness. However, the wind crosses the surface at an oblique angle to the template profiles. The geometric mean of the RMS height for both profiles gives a better "3-D" characterization of the surface, which correlates well ($r = 0.84$) with aerodynamic roughness estimates (Fig. 10). There is a similarly good correlation ($r = 0.91$) between mean RMS height for the laser profiles at each site and aerodynamic roughness (Fig. 10). The RMS height derived from these data are an order of magnitude less than those for the template data. This suggests that the laser data provide a measure of particle roughness at alluvial fan sites and microtopography at the smooth site (Confidence Mill playa).

Greeley and Iversen [4] have suggested that aerodynamic roughness increases from 1/30 the diameter of surface particles when they are closely packed to a maximum of 1/8 particle diameter when the particle spacing is about twice the diameter. Further increases

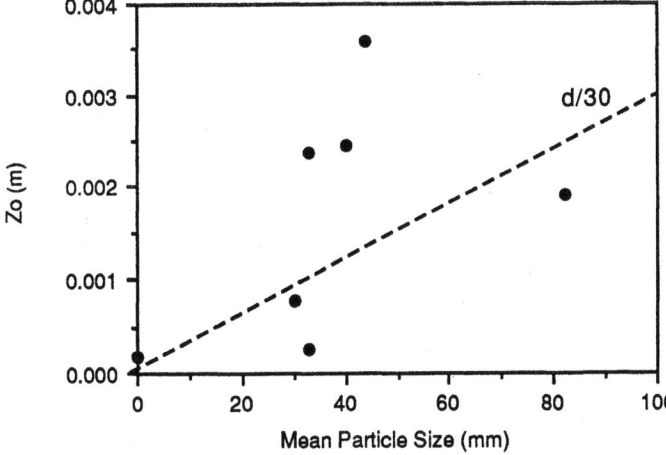

Fig. 8. Relationship between aerodynamic roughness (southerly wind data) and estimated mean particle size at study sites. $d/30$ relationship shown for comparison

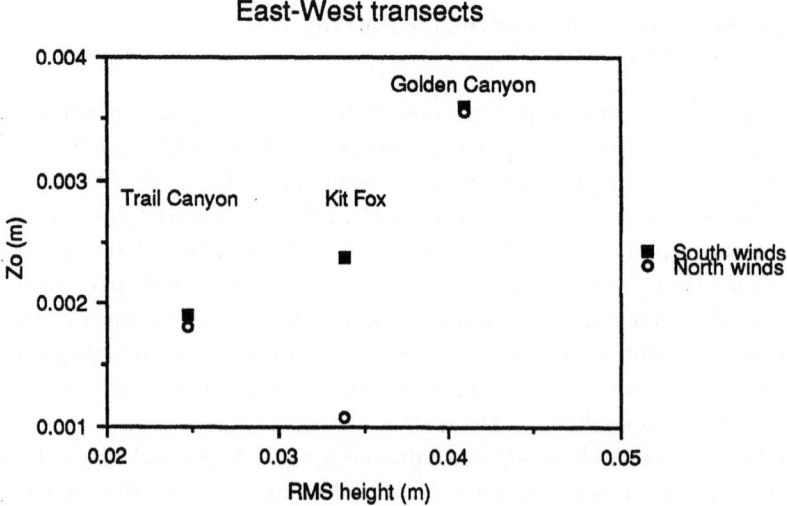

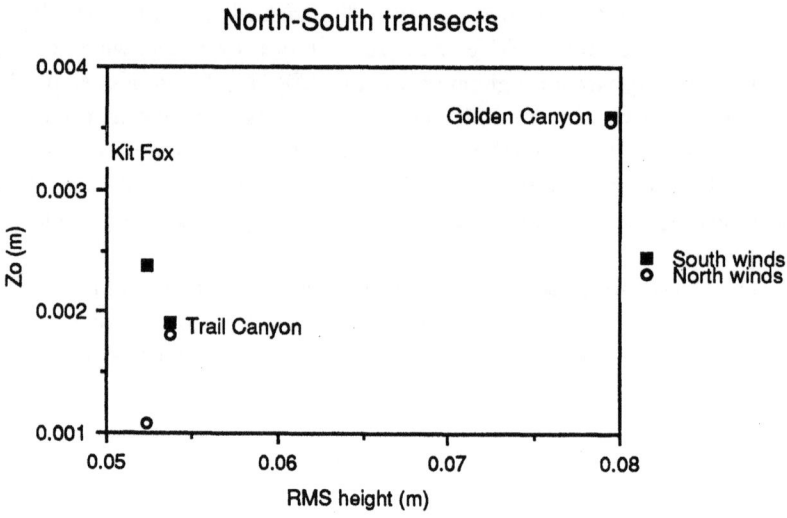

Fig. 9. Relationship between aerodynamic roughness and RMS height (unfiltered data) for both E−W and N−S transects

in particle spacing result in a decrease in aerodynamic roughness. The data from Death Valley sites suggest that at sites where the surface consists of coarse particles (gravel, cobbles) in a fine matrix, aerodynamic roughness is $\sim 1/15$ to $1/20$ the mean template RMS height, and $\sim 1/4$ the mean laser RMS height. This is true for all sites except Stovepipe Wells where the aerodynamic roughness is $\sim 1/20$ the laser RMS height. At "smooth" sites such as Confidence Mill playa, aerodynamic roughness is $1/30-1/40$ the mean laser RMS height. These preliminary results suggest that the surfaces are behaving in different ways with respect to the wind, depending on the size and arrangement of particles on the surface.

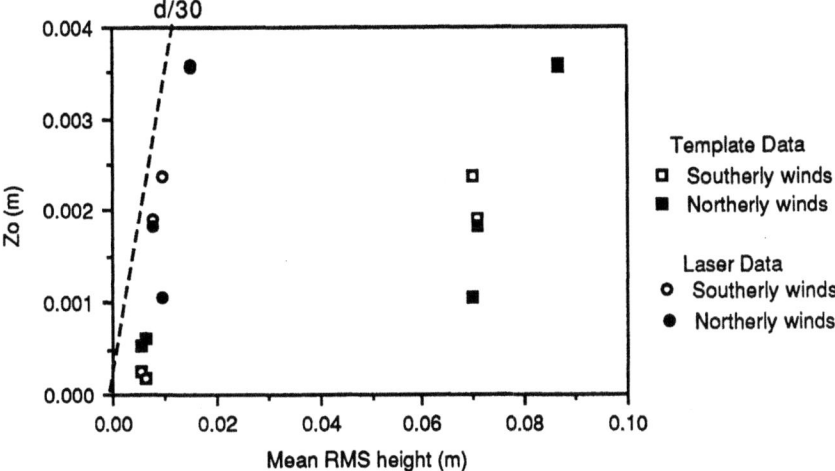

Fig. 10. Relationship between mean RMS height derived from template and laser data and aerodynamic roughness for all study sites. $d/30$ relationship shown for comparison

6 Conclusions

These studies show clearly that the aerodynamic roughness of an unvegetated surface is a function of both its microtopographic and particle roughness characteristics. There is a good relationship between unfiltered RMS height and aerodynamic roughness, suggesting that the RMS height of the surface is a good index of its overall topographic roughness characteristics. Further analyses of the existing data and additional field experiments are however required to understand the contribution of different scales of topographic and particle roughness to the aerodynamic roughness estimates, and to study the effects of varying roughness characteristics on sediment transport.

Acknowledgements

This resarch was conducted in Death Valley National Monument with permission from the Superintendent. We thank the National Park Service for their assistance and support. Special thanks are due G. Beardmore for technical assistance and field support, D. Ball for photographic services and advice, and J. Lancaster for the fieldwork. This investigation was supported by the National Aeronautic and Space Agency through Ames Research Center Grant NCC 2-346 and Jet Propulsion Laboratory Contract 88-04070. Rasmussen thanks the Danish Natural Science Research Council and the NATO Science Fellowship Programme for support.

References

[1] Greeley, R., Iversen, J. D.: Measurements of wind friction speeds over lava surfaces and assessment of sediment transport. Geophys. Res. Lett. 14, 925−928 (1987).

[2] Greeley, R., Lancaster, N., Sullivan, R. J., Saunders, R. S., Theilig, E., Wall, S., Dobrovolskis, A., White, B. R., Iversen, J. D.: A relationship between radar backscatter and aerodynamic roughness: preliminary results. Geophys. Res. Lett. 15, 565−568 (1988).

[3] Lyles, L., Schrandt, R. L., Schneidler, N. F.: How aerodynamic roughness elements control sand movement. Trans. Am. Soc. Agr. Eng. 17, 134−139 (1974).

[4] Greeley, R., Iversen, J. D.: Wind as a geological process, p. 333. Cambridge: Cambridge University Press 1985.

[5] Gillette, D. A., Adams, J., Endo, A., Smith, D.: Threshold velocities for the input of soil particles into the air by desert soils. J. Geophys. Res. **85**, 5621—5630 (1980).

[6] Fleagle, R. G., Businger, J. A.: An introduction to atmospheric physics, p. 432. New York: Academic Press 1980.

[7] Höström, U.: Non-dimensional wind and temperature profiles in the atmospheric surface layer: a re-evaluation. Boundary Layer Meteorol. **42**, 55—78 (1988).

[8] Rasmussen, K. R. Some aspects of flow over noastal sand dunes Proc. Roy. Soc. Edinburgh. Series B, (1990).

[9] Garratt, J. R.: Flux profile relations above tall vegetation Quart. J. Roy. Met. Soc. **104**, 199—211 (1978).

[10] Bagnold, R. A.: The physics of blown sand desert dunes, p. 265. London: Chapman and Hall 1941.

Authors' address: N. Lancaster, R. Greeley, and K. R. Rasmussen, Department of Geology, Arizona State University, Tempe, AZ 85287-1404, U.S.A.

Acta Mechanica (1991) [Suppl] 2: 103—112

The threshold friction velocities and soil flux rates of selected soils in south-west New South Wales, Australia

J. F. Leys, Buronga, Australia

Summary. A portable wind erosion tunnel has been used to measure the wind erodibility of nine soil types with a range of surface textures under two treatments (bare uncultivated and bare cultivated) in western New South Wales. The erodibility has been characterised by the soil flux function $Q(u_*)$, where Q is the streamwise soil flux (measured with a modified Bagnold soil trap) and u_* the friction velocity (obtained by fitting the logarithmic wind profile law to a wind profile measured in the tunnel). Threshold friction velocities u_{*t} were also observed for the range of surfaces. These data represent the only Australian tests published to date and supplement the American measurements of semi-arid soils by Gillette [1], [2], [3].

Averaging over 10 replicate plots for each surface type was necessary to smooth the large scatter in Q and smaller scatter in u_*. The Q values spanned three decades of magnitude. Soils with a sandy loam surface texture were the boundary between the highly erodible sand and the basically nonerodible clay. Cultivation increased the erodibility of the majority of soils by about a factor of 10, but decreased the erodibility of the clay. The function $Q(u_*)$ is well described by the Owen [4] soil flux equation. Threshold friction velocity decreased as soil texture became sandier. In comparison with work of Gillette [3], Australian soils have lower u_{*t} values, which is most likely due to higher sand and lower silt contents.

1 Introduction

Wind erosion affects agricultural productivity and the general community over much of Australia, in both the short and the long term. The most obvious local and immediate effect is loss of topsoil, rich in nutrients and organic materials, which leads directly to productivity loss [5], [6]. It is also generally accepted that wind erosion involves the selective winnowing of clay and silt out of the soil, leading in time to reduced water and nutrient-holding capacity, decreased aggregation and increased susceptibility to further wind erosion [7]. The nonlocal effects of wind erosion include air pollution from dust and haze, and the deposition of dust and sand over the landscape. Nonlocal costs (principally concerned with cleanup) may dwarf local on-site costs [8].

The traditional farming system practiced in the mallee areas of South Australia, Victoria and New South Wales (N.S.W.) used disc ploughs and scarifiers to maintain a bare (weed free) fallow for nine months prior to sowing. This practice of "long" fallowing aims to reduce cereal disease, mineralise nitrogen and conserve moisture. A three year rotation of volunteer pasture/fallow/wheat with grazing of the pasture and stubble, was the normal practice until five years ago [9].

Wind erosion and soil drift was inevitable with this bare fallow system because the soil

Table 1. Measured and derived properties of sites and treatments: Surface soil texture;
(a) — observed threshold friction velocity u_{*t} cm s^{-1};
(b) — observed u_{*t} cm s^{-1} from Gillette [3];
f_{agg} — percentage dry aggreagation $f_{agg} > 0.84$ mm;
Q — soil flux Q g m^{-1} s^{-1}

Treatment →		Cultivated[1]				Uncultivated[2]			
Site	Texture	u_{*t} (a)	u_{*t} (b)	f_{agg}	Q	u_{*t} (a)	u_{*t} (b)	f_{agg}	Q
H	sand	30	25—37	0.05	123.70	31[4]	25—37	0.26	106.59
A	sand	17	25—37	0.16	115.90	38	90	0.45	14.08
D	sandy loam	79[3]	> 78	0.40	5.56	65	63—336	0.77	0.33
E	sandy loam	23	22—44	0.25	17.72	50	> 63	0.62	5.06
G	loam	46	67—91	0.37	6.31	64	> 100	0.78	2.34
B	sandy clay loam	19	n.a.	0.37	3.79	42	n.a.	0.71	0.58
I	loam fine sandy	54[3]	n.a.	0.67	1.26	60	n.a.	0.93	0.59
F	clay loam	73[3]	> 71	0.69	1.02	75	> 100	0.91	0.12
C	clay	53[3]	> 150	0.93	0.18	44[5]	> 63	0.97	0.35

Notes: 1 — All soils with loose surfaces (exceptions noted).
2 — All soils had crusted surfaces (exceptions noted).
3 — These soils had a cloddy surface.
4 — Although this soil was uncultivated it had a loose surface.
5 — This soil had a sandy loam veneer overlying the clay.
n.a. — Not available

was left devoid of vegetation over the summer fallow months when strong, hot, dry winds occur. Good wheat prices in the late 1970's and early 1980's saw an expansion in the area cropped in the far south-west of N.S.W. [10]. This brought increased awareness of the erosion problems which occurred with the traditional long fallow system.

Two measures which have helped to combat agricultural wind erosion in recent years are avoidance of highly wind erodible land for cropping and the use of stubble retention farming systems. The decision as to what is "capable" of being cleared and cropped is the overall aim of this research. The Soil Conservation Service of N.S.W. aims to use the wind tunnel to provide objective soil erodibility information to delineate the highly erodible soil types/landscapes.

"Erodibility" is a property of a particular soil, defined as the susceptibility of the soil particles to detachment and transport by an erosive agent, in this case wind [11].

This paper continues the work of Leys and Raupach [12] that investigated the wind erodibility of three soil types and assessed the soil flux as a function of friction velocity $Q(u_*)$. They found that Owen's sand transport (Equation 3) was satisfactory for all surfaces tested except those with high dust emissions. This paper further describes investigations of the wind erodibility of an additional six soil types, (giving a total of nine) ranging in surface texture from sand to clay, under two surface treatments (bare uncultivated and bare cultivated). This will allow further assessment of the soil flux function $Q(u_*)$ and determine if $Q(u_*)$ continues to fail on soils with high dust emissions.

The quantity u_* (with units m s^{-1}) is a convenient measure of the shear stress τ exerted by the wind on the ground surface, and is defined by $u_* = (\tau/\varrho)^{1/2}$ (ϱ being the air density). The importance of u_* is that the shear stress τ on the ground surface equals the flux density of streamwise momentum from the air to the ground; this momentum transfer is the means by which soil movement is initiated and sustained, so u_* is the wind velocity scale most directly related to wind erosion. The function $Q(u_*)$ (for a particular soil, in a given surface state defined by cultivation treatment, vegetation, moisture status and so on) can be thought of as defining the erodibility of that surface.

The aim of this paper is to quantify the erodibility of each surface (nine soils by two treatments) by determining experimentally the relationship between wind velocity and soil transport, and to interpret the observed erodibilities in terms of soil type and treatment.

2 Methods and results

Full descriptions of the methods are available elsewhere [12] but are briefly summarized here. The measurements were made on nine soils with different surface textures (numbered A to I) (Table 1). Two treatments were investigated for each surface texture: bare uncultivated (denoted n) and bare cultivated (denoted c). The wind velocity and soil transport measurements were made with the portable wind erosion tunnel.

The methodology was to run the tunnel for 60 s periods at each of a range of wind velocities (3 to 14 m s^{-1} at $z = 20$ mm). Soil was collected via a Bagnold sand trap connected to a vacuum system in which the soil was accumulated in filter bags for subsequent weighing to determine soil flux Q. Dynamic pressures (p_d) were measured from an array of Pitot-static tubes for the subsequent determination of mean wind velocity $\bar{u}$ and thence the friction velocity u_*. Ten replicates for each treatment were measured for each surface texture.

The procedure for finding u_* and z_0 was to fit the logarithmic wind profile law (1) to the measured profile of $\bar{u}(z)$.

$$\bar{u}(z) = \lambda(u_*/k) \ln (z/z_0) \tag{1}$$

where z_0 is the roughness length and k is the von Karman constant (taken as 0.4).

For each surface texture, additional soil measurements were made as follows. A measure of soil structure (percentage dry aggregation with diameter greater than 0.85 mm, denoted f_{agg}, was determined by gentle hand sieving using a 0.85 mm sieve [13]. Soil particle size analysis was carried out on 22 systematic samples from each site, using disturbed samples from depth $0-50$ mm.

3 Discussion

There are two main sources of scatter in the data. Firstly, field variability caused Q values from the 10 individual plots for each surface (An, Ac, ...) to vary substantially, with a standard deviation σ roughly proportional to the mean value of Q itself. Secondly, the Pitot-static values of $\bar{u}$ exhibited scatter between plots because of the short averaging time (60 s), which introduced some turbulent fluctuations into $\bar{u}$. To remove both sources of scatter, the data from the 10 replicate plots for each surface were averaged together on the assumption that they were members of the same statistical ensemble. This made possible the calculation of average values of Q, $\bar{u}$ (at height 50, 100 and 200 mm), u_* and z_0, for all surfaces. Indications of the scatter of individual-plot data about these averaged values are given later.

Wind profiles, friction velocities and roughness lengths

The log law (1) was used to plot the wind velocity profile and calculate u_*. Figure 1 shows $\bar{u}(z)$ plotted against $\ln(z)$, for surface Ac.

This procedure involves several assumptions and neglects some possible errors, leading overall to uncertainties in the order 10 to 20% in the calculated values for u_*. These are discussed at length elsewhere [14]. In summary, the main difficulties are:

(a) While 0.40 is the generally accepted value for k, direct measurements in this tunnel [14] found a value which was lower by 5 to 10%. Despite this, $k = 0.4$ was used here.

(b) There is some uncertainty about the level of the aerodynamic height origin, $z = 0$. The resulting uncertainty of about 1 cm in z implies an uncertainty of about 5% in u_*.

(c) Equation (1) is inapplicable within the saltation layer [15]. It was observed that most of the saltating soil within the tunnel was below the lowest $\bar{u}$ measurement level, $z = 5$ cm, so the resulting error is expected to be small.

(d) There is scatter in $\bar{u}$ because of the short measurement period. Data for surface Ac is representative of the scatter in $\bar{u}$ for the other surfaces. Figure 1 indicates by error bars (for the lowest and highest speeds only) the uncertainty in $\bar{u}$ after averaging over 10 plots for surface Ac. The uncertainty is small, but is highest near the ground where the turbulence intensity is large [14]. The scatter in $\bar{u}$ probably imposes an uncertainty of about 5% in u_*.

The results for u_* and z_0, obtained by applying Equation (1) to the measured wind profiles over each surface at each speed, show two features. Firstly, and not surprisingly, roughness lengths z_0 are lower for uncultivated surfaces than their cultivated counterparts. The roughest surfaces were for those soils with higher clay contents ($> 18\%$) and a

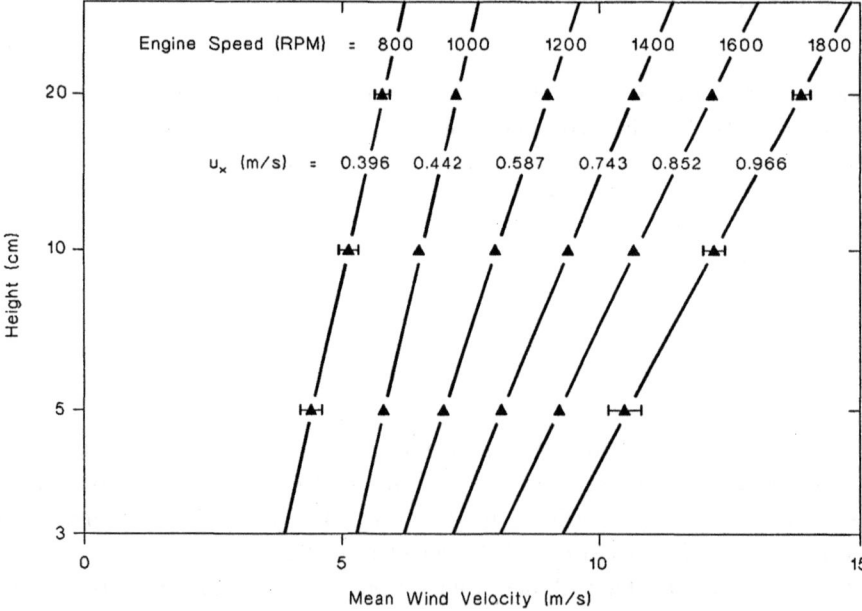

Fig. 1. Mean wind velocity $\bar{u}$ plotted against the logarithmic height $\ln(z)$ for the determination of u_* and z_0, for surface Ac. The six straight lines represent fitted logarithmic profiles for each of six engine speeds. Points are means over 10 replicate plots. For the lowest and highest engine speeds, the error bars show uncertainty in the mean [the sample standard deviation divided by $(10-1)^{-1/2} = 3$]

massive structure, such as Cc, Fc, Gc and Ic. Although Bc had 22% clay it was not a massively structured soil and produced less large clod on the surface. The smoothest surface is Bn which exhibited a strong crust.

Secondly, for each surface type, the ratio $u_*/\bar{u}(z)$ is independent of friction velocity to within $\pm$ 15%. The exceptions are Cn, Fn and Gn surfaces which have $u_*/\bar{u}(z)$ ratios of 40, 48 and 19% respectively. Each of these exceptions had the greatest difference in $u_*/\bar{u}(z)$ between the lowest and the highest wind velocity and each of these soils have a high clay content.

Since $u_*/\bar{u}(z)$ determines z_0 through the relationship $z_0 = z \exp(-k\bar{u}(z)/u_*)$, it follows that z_0 is likewise independent of friction velocity. It is important to note that z_0 is very sensitive to $u_*/\bar{u}(z)$, particularly over the smoother surfaces such as Bn. Over these surfaces, small changes in $u_*/\bar{u}(z)$ (of order 10%) produce large apparent changes in z_0 which are not significant in reality. Bearing this in mind, the conclusion is that the drag coefficient and the surface roughness are essentially independent of friction velocity for all surfaces in our measurements.

For the majority of surfaces, z_0 shows no sign of increasing with friction velocity. The exceptions are the sandy loam and sand surfaces of Ec, Hc and Hn which exhibit marginal increases. This is in contrast to the behaviour of z_0 observed over strongly saltating surfaces, where z_0 increases with wind velocity [15]. An approximate description of this increase is

$$z_0 = c_0 u_*^2/g \tag{2}$$

with g being the acceleration due to gravity ($g = 9.8 \mathrm{~m~s^{-2}}$) and c_0 ($= 0.016$) a dimensionless constant [15] [16]. Equation (2) does not apply to our data because it is valid only when $u_* \gg u_{*t}$ (where u_{*t} is the threshold friction velocity), a condition implying much stronger saltation than was observed.

Two reasons can be offered as to why strong saltation was not observed on most surfaces. Firstly, there was insufficient material available for entrainment and therefore the potential transport capacity of the wind was not reached. Secondly, the tunnel length is too short for a fully developed saltation layer to develop.

Streamwise soil fluxes

Qualitatively, the highest soil transport rates Q for the uncultivated treatments occur over the sands (sites H, A and E) and is lowest over the clays (F and I), with loams (C, G, and B) yielding intermediate transport rates. The one exception is the sandy loam (D), which has a transport rate similar to a loam, although its surface texture is much sandier (Table 1). This trend is the same for the cultivated treatment. The overall range of Q spans three decades of magnitude. The effect of cultivation on all soils [except the clay (C)] is to greatly increase Q. Cultivation of the clay (C) decreases Q from a low to an even lower value.

These trends are broadly consistent with the observed soil structures and responses to cultivation. Firstly, the increase in clay and silt fractions through the texture sequence from sand to clay progressively reduces Q by increasing the proportion of dry aggregation at the soil surface. The one exception is the sandy loam (D) which for both treatments has a Q value similar to the loams. Table 1 gives the percentage of aggregation f_{agg} of dry aggregates (larger than 0.85 mm diameter) and the soil flux Q for each soil, confirming this interpretation.

Secondly, the effect of cultivation on the soil, except for the clay (C), is to pulverise it and greatly reduce the percentage of dry aggregates (Table 1), leading to large increases in

Q relative to the uncultivated state. The small reduction in Q for Cc compared to Cn is explained by the presence of a thin layer of sandy loam on the surface of the uncultivated clay soil, which provided a small amount of erodible material. After cultivation, this was incorporated into the soil and highly structured material (with a percentage aggregation well over 90%) dominated the surface of Cc, thus reducing the amount of erodible material at the surface.

When σ_Q (the sample standard deviation in Q) is plotted against Q for the cultivated and uncultivated treatments, the plot-to-plot scatter in Q is very high. Approximately, σ_Q is of the same order as Q, with the exception of the sandy loam (D) which has a lower σ_Q than Q. This implies that the standard error in the mean values of Q is about $(10 - 1)^{1/2} Q$ $= Q/3$, an error small enough to permit the above qualitative conclusions to be drawn with confidence. However, the fact that σ_Q/Q approximately equals 1 meant that averaging over many replicate plots was required.

To go beyond these qualitative statements, it is necessary to describe the data mathematically with a soil transport equation which specifies the soil flux function $Q(u_*)$. Many sand/soil transport equations have been developed over the years and are summarised on page 100 of Greeley and Iversen [17]. From these, Owen's sand transport equation [4] was selected because it had successfully been used by Gillette and Stockton [18] to fit soil transport rates and because it most unambiguously described the theoretical relationship between sand transport and u_*.

The sand transport equation was not expected to fit the Q data from all the surfaces because many of the assumptions made by Owen [4] could not be met. These included assumptions that potential Q was achieved for a given u_* (above u_{*t}) and the transported material was of uniform size. Despite known limitations, the sand transport equation was used to evaluate its reliability in predicting the function $Q(u_*)$ over a range of soil textures and to distinguish between those surfaces that behaved like sands (highly erodible surfaces) and those that did not.

Using the methodology described in Leys and Raupach [12]. It is assumed that $Q(u_*)$ is of the form predicted by the linearized theory of Owen [4] for the transport of sand of uniform grain size by saltation (see his Equation (41)):

$$Q = \frac{c\varrho u_*^3}{g}\,(1 - u_{*t}^2/u_*^2), \qquad u_* > u_{*t} \tag{3a}$$

$$Q = 0, \qquad u_* = < u_* \tag{3b}$$

where c is a dimensionless parameter and u_{*t} is the threshold friction velocity.

It is considered that c and u_* are parameters which can be determined for each surface type by fitting Equation (3) to data of the form (u_{*i}, Q_i), $i = 1, ..., n$, where n ($= 5$ or 6, sites A to C; or 8, sites D to I) is the number of speed settings available for that surface type. The analysis was done by a least-squares procedure summarised elsewhere [12].

The description of the data by Equation (3) is, on the whole, good. The resulting fits of the equation for a sandy loam and loam fine sandy surface textures are shown in Figures 2a and 2b, highlighting the good fit with sandy soils and poor fit with clayey soils.

Difficulties occur with the fitting only when the soil flux is very low ($Q < 0.1\ \mathrm{g\ m^{-1}\ s^{-1}}$) or dust emissions are high. In these cases, the soil fluxes are likely to be dominated by emissions of fugitive dust and other loose material from the surface, which is transported mainly in suspension. Such transport cannot be expected to obey a sand transport equation formulated for particle movement by saltation [4]. It is therefore not surprising that

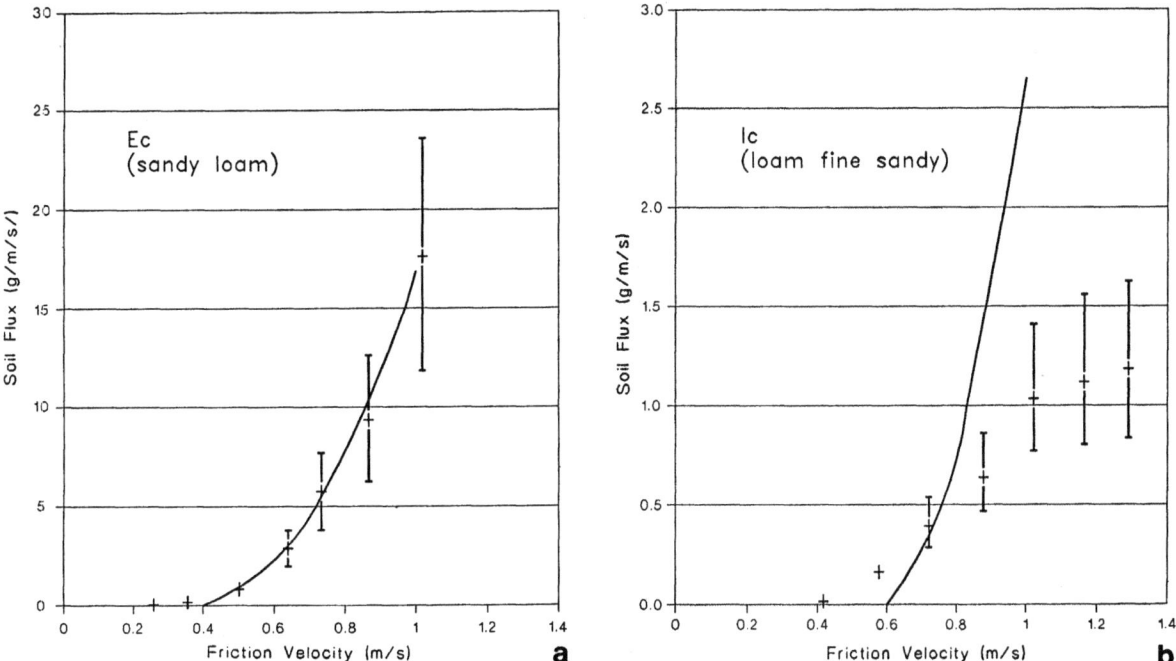

Fig. 2. Fits of sand transport equation specifying $Q(u_*)$, Equation (3), to soil transport data for sites *Ec* (Fig. 2a) and *Ic* (Fig. 2b). The fitting procedure works well for the sandy soils where saltation dominates the ransport process but fails for the clayey soils which exhibit high dust emission or ow soil flux Q values

small, but detectable, soil fluxes Q are observed on the surfaces of soils with high clay contents (*Bn, Cc, Fn, Ic* and *In*) at low u_* values. In the cases of *Cc, Fc* and *Ic*, these emissions dominate the data to the extent that the fitted u_{*t} value is zero, which indicates no more than that Equation (3) is inapplicable to these surfaces with dust emissions.

The theory leading to Equation (3), which strictly applies only to pure saltation of sand of uniform grain size, carries the implicit assumption that c is a constant of order 1. The idealisations of the theory are probably reasonable for the sandy soils (*A, E* and *H*), where c is in an order of magnitude of 1. However, for the remaining soils with higher clay contents, complications include substantial nonuniformity in particle size distributions and the presence of nonerodible aggregates which shelter the erodible surface. These two factors interact, because as u_* increases, progressively larger particles are mobilised and are thereby transfered from the nonerodible fraction to the erodible fraction. A third factor is the thin fragile crust on the uncultivated surfaces which inhibited particle movement at low friction velocities. All non cultivated surfaces exhibited some crusting (with the exception of *Hn*).

The combined effects of nonuniform size distribution, sheltering by nonerodible aggregates, and crusting lead to reduction of the constant c in Equation (3) to values much less than 1. Given the real-world complications neglected in the theory underlying Equation (3), its adequacy in describing most of the soil flux data is pleasantly surprising.

Threshold friction velocities

The threshold friction velocity u_{*t} was measured when $Q > 0.1$ g m^{-1} s^{-1} over a one minute period at a constant u_*. The results are shown in Table 1. The u_{*t} values predicted from equation 3 were not used.

The observed u_{*t} values from the nine soils were compared with the work of Gillette [3] who measured the u_{*t} of a large range of soils in agricultural and semi-arid areas [1], [2]. In general there is close agreement between the American and Australian data sets for the cultivated soils. However, for the uncultivated surfaces the Australian u_{*t} values are generally lower. It is the author's opinion that this is because Australian soils have a higher sand and lower silt contents than comparable American soils of the same texture class. These differences result in lower crust strength and more loose sand grains (lag material after past erosion events) on the surface. These sand grains are then remobilised during erosion events at lower u_{*t} compared to the crusted surfaces they rest upon.

As the percentage of fines increases (i.e. increasing clay + silt content) u_{*t} also increases. The relationship is stronger for the cultivated surfaces than the uncultivated ones. This is probably because the fines are having less effect on the uncultivated surface than other factors such as crusts and availability of erodible material on the surface (a function of past erosion history).

For the cultivated surfaces, there is a trend between percentage fines and u_{*t} (regression coefficient of $r^2 = 0.21$) which takes the form

$$u_{*t} = \big((\%\text{fines})\,0.633 + 26.326\big). \tag{4}$$

The uncultivated surface has a weaker trend and while both the regression lines are not significant, a trend does appear likely with u_{*t} increasing as percentage fines increases.

4 Conclusions

Measurements, made with a portable wind erosion tunnel, have been presented of the wind erodibility of nine common soil surface textures (numbered A to I) in western New South Wales. The soil textures ranged from sand to a clay, and were subject to two treatments, bare uncultivated (n) and cultivated (c). As a quantitative measure of erodibility the soil flux function $Q(u_*)$ was used, where Q is the streamwise soil flux and u_* the friction velocity. The main conclusions are:

(1) Mean wind profiles in the tunnel, measured with Pitot-static tubes at two or three heights (z) between 50 and 200 mm, are well described by a logarithmic wind profile law (Equation 2). The fitted profile yields satisfactory measures of the friction velocity u_* and roughness length z_0. For all surfaces studied, $u_*/\bar{u}(z)$ (at a height z of 50 mm) was independent of $\bar{u}$ to within $\pm 15\%$, and z_0 did not increase with $\bar{u}$.

(2) The measured streamwise soil flux Q spanned three decades of magnitude, from the highly erodible sand to the essentially nonerodible clay. Cultivation increased Q for all soil textures by about a factor of 10 (range of 2 to 16) because of the associated decrease in soil aggregation. On the other hand, Q for clay (site C) was decreased by cultivation due to the high soil aggregation and the protective effect of nonerodible clods.

(3) The sample standard deviation in Q from plot to plot, σ_Q, is of the same order as Q. Substantial averaging over replicate plots was necessary. Ten replicates were used.

(4) Owen's [4] sand transport equation (Equation 3) was previously tested [12], on soils A, B, and C and was found to provide a satisfactory form for the soil flux function $Q(u_*)$. Testing of a further six soils shows that Equation 3 is suitable for the soil flux function $Q(u_*)$ with the exception of freshly cultivated soils with high dust emissions at low u_* giving unrealistic fits to the data. Equation 3 is also inappropriate for the calculation of

u_{*t} because one of its implicit assumptions is that c remain in the order of unity. This was not achieved in the data fitting procedure.

(5) The u_{*t} increases as the percentage of fines (clay + silt) of a soil increases, ranging from 17 to 73 cm s^{-1} for cultivated surfaces. The same trend applies for uncultivated surfaces but the trend is less pronounced due to the effects of surface crusts and past erosion histories of the sites. The u_{*t} values range from 31 to 75 cm s^{-1}.

(6) The u_{*t} values from the Australian soils are similar to those reported by Gillette [3] for cultivated soils but are generally less (by a factor of two) compared to American uncultivated soils. It is the author's opinion that this is due to the high sand and low silt contents of Australian soils compared to American soils of similar surface texture.

(7) Of all the soils tested, the sandy loam (D) consistently yielded anomalous results to the other similarly textured soils. From experience, this soil erodes less than other soils with similar surface textures in the area. Further detailed particle size work is being undertaken to identify if it has different properties to the other tested soils.

In conclusion, those soils which are highly erodible can be delineated and recommendations as to their capability can be made for south-west N.S.W. There are three groups of erodibility, high, moderate and low. The sands (A and H) and the sandy loam (E) soils are highly erodible, especially if cultivated. They would not be recommended for clearing or cropping but would be suitable for extensive grazing at low stocking rates. The sandy clay loam (B), sandy loam (D), loam (G) and loam fine sandy (I) are moderately erodible and are capable of supporting some clearing, cropping with stubble retention farming, conservative grazing and sowing to introduced pastures. The clay (C) and clay loam (F) have low erodibility and are capable of supporting clearing and cropping with no special conservation practices.

Acknowledgements

This research has been partially supported by a grant from the Rural Credits Development Fund of the Reserve Bank of Australia to the Soil Conservation Service of New South Wales.

The author acknowledges the contribution of Dr. M. Raupach for the development of the mathematical fitting of the data and the helpful comments of Prof. B. Willetts and Dr. D. Gillette on the earlier draft.

The assistance of Mr. D. Cosgrove with collection and development of the wind tunnel is gratefully acknowledged. The author would like to thank Messrs. W. S. Semple, P. J. Walker and C. A. Booth for their continual encouragement, and the landholders on whose properties the research was undertaken.

References

[1] Gillette, D., Adams, J., Endo, A., Smith, D., Kihl, R.,: Threshold velocities for the input of soil particles into air by desert soils. J. Geophys. Res. 85, 5621−5630 (1980).

[2] Gillette, D., Adams, J., Muhs, D., Kihl, R.: Threshold friction velocities and rupture moduli for crusted desert soil for the input of soil particles into air. J. Geophys. Res. 87, 9003−9015 (1982).

[3] Gillette, D.: Threshold friction velocities for dust production for agricultural soils. J. Geophys. Res. 93, 12, 645−662 (1988).

[4] Owen, P. R.: Saltation of uniform grains in air. J. Fluid Mech. 20, 225−242 (1964).

[5] Chepil, W. S., Woodruff, N. P.: The physics of wind erosion and its control. Adv. Agron. 15, 211−302 (1963).

[6] Lyles, L., Tartako, J.: Wind erosion effects on soil texture and organic matter. J. Soil Water Cons. **41**, 191—193 (1986).

[7] Fryrear, D. W.: Long term effect of erosion and cropping on soil productivity. In: Desert dust: origin, characteristics and effect on man. (T. L. Pewe, ed.). Boulder, CO: Geological Society of America, pp. 253—259 (1981).

[8] Huszar, P. C., Piper, S. L.: Estimating off-site costs of wind erosion in New Mexico. J. Soil Water Cons. **41**, 414—417 (1984).

[9] Leys, J. F.: Blow or grow? A soil conservationists view to cropping mallee soils. In: The Mallee lands: a conservation perspective. (Noble, J. C., Joss, P. G., Jones, G. K., eds.) CSIRO. Melbourne (in press).

[10] Eldridge, D. J., Semple, W. S.: Cropping in marginal south-western New South Wales. J. Soil Cons. Serv. N. S. W. **38**, 65—71, (1982).

[11] Houghton, P. D., Charman, P. E. V.: Glossary of terms used in soil conservation, p. 147. Sydney: Soil Conservation Service of New South Wales, (1986).

[12] Leys, J. F., Raupach, M. R.: Soil flux measurements using a portable wind erosion tunnel. Aust. J. Soil Res. **29** (4) (in press).

[13] Semple, W. S., Leys, J. F.: The measurement of two factors affecting the soil's susceptibility to wind erosion in far south-west New South Wales: soil roughness and proportion of nonerodible aggregates. Soil Cons. Serv. New South Wales Tech. Bull. **28**, p. 20 (1987).

[14] Raupach, M. R., Leys, J. F.: Aerodynamics of a portable wind erosion tunnel for measuring soil erodibility by wind. Aust. J. Soil Sci. **28**, 177—192 (1990).

[15] Bagnold, R. A.: The physics of blown sand and desert dunes. p. 265. London: Methuen (1941).

[16] Chamberlain, A. C.: Roughness length of sand, sea and snow. Boundary-Layer Meteorol. **25**, 405—409 (1983).

[17] Greeley, R., Iversen, J. D.: Wind as a geological process. Cambridge: Cambridge University Press (1985).

[18] Gillette, D. A., Stockton, P. H.: The effect of nonerodible particles on wind erosion of erodible surfaces. J. Geophys. Res. **94**, 12885—12893 (1989).

Author's address: J. F. Leys, Soil Conservation Service of New South Wales, P.O. Box 7, Buronga N.S.W. 2648, Australia.

Acta Mechanica (1991) [Suppl] 2: 113—130
© by Springer-Verlag 1991

Wind degradation on the sandy soils of the Sahel of Mali and Niger and its part in desertification

M. Mainguet and M. C. Chemin, Reims, France

Summary. From South Mauritania to Sudan between the 150 and 600 mm isohyets the sub-Saharan semi-arid lands are covered by a fixed sand sea. Two areas are discussed in this paper: in Niger west and east of the river Niger bend from N'Guigmi to Niamey and in Mali leeward of the Niger's inland delta. The aim is 1) to describe the morphometric data of the sand accumulation, 2) to understand the actual aeolian removal of particles, 3) to analyse the effects of wind erosion in desertification.

The results obtained show that the deposits south of Sahara accumulated on a surface of 600 km are due to the same Wind Action System (WAS). The grain size is constantly and globally decreasing in the Harmattan wind direction. The wind abrasion on the sand particles is increasing in the same direction. The grain size 400 μm is a threshold: the upper limit of actual, average aeolian deflation.

This area, which previously had a positive sediment balance, during the last drought combined with agricultural overuse became an area of sand and dust removal with a negative sediment balance.

Introduction

From Mauritania to Sudan the sub-Saharan drylands are partly covered by aeolian sand sheets with a total area of 10^6 km². These deposits are red coloured, stabilized dunes from an old sand sea accumulated during a dryer period in the geological recent history, probably the 20.000 BP drought (Petit-Maire [22]). When the conditions became more humid a paleo-soil formation and a vegetative cover stabilized the aeolian deposits. If the vegetation is removed the dunes will easily become unstable again.

Our aim is: First to describe the location of and characterize the deposits in the sand accumulations in Mali and Niger. Secondly, to try to understand how land degradation occurs in the Sahelian ecosystem. Thus our aim is also to throw light on the problems mentioned on the Wind Erosion Conference [3] at Lubbock, Texas. The introduction of the conference report says: "We need to know more about sorting of soil material during the erosion process and the qualitative effect on soil productivity." A final aim is to see the effect of wind erosion in desertification.

1 Description of the area

1.1 Morphological aspects of the sub-Saharan stabilized sand sea

The southern active sand sheet of the Sahara Desert has extended in the Sahel (between the 150 and 600 mm isohyets) as a stabilized dune belt stretching from Mauritania to Senegal through Mali, north of Burkina Faso, Niger, Chad (at the latitude of the lake Chad)

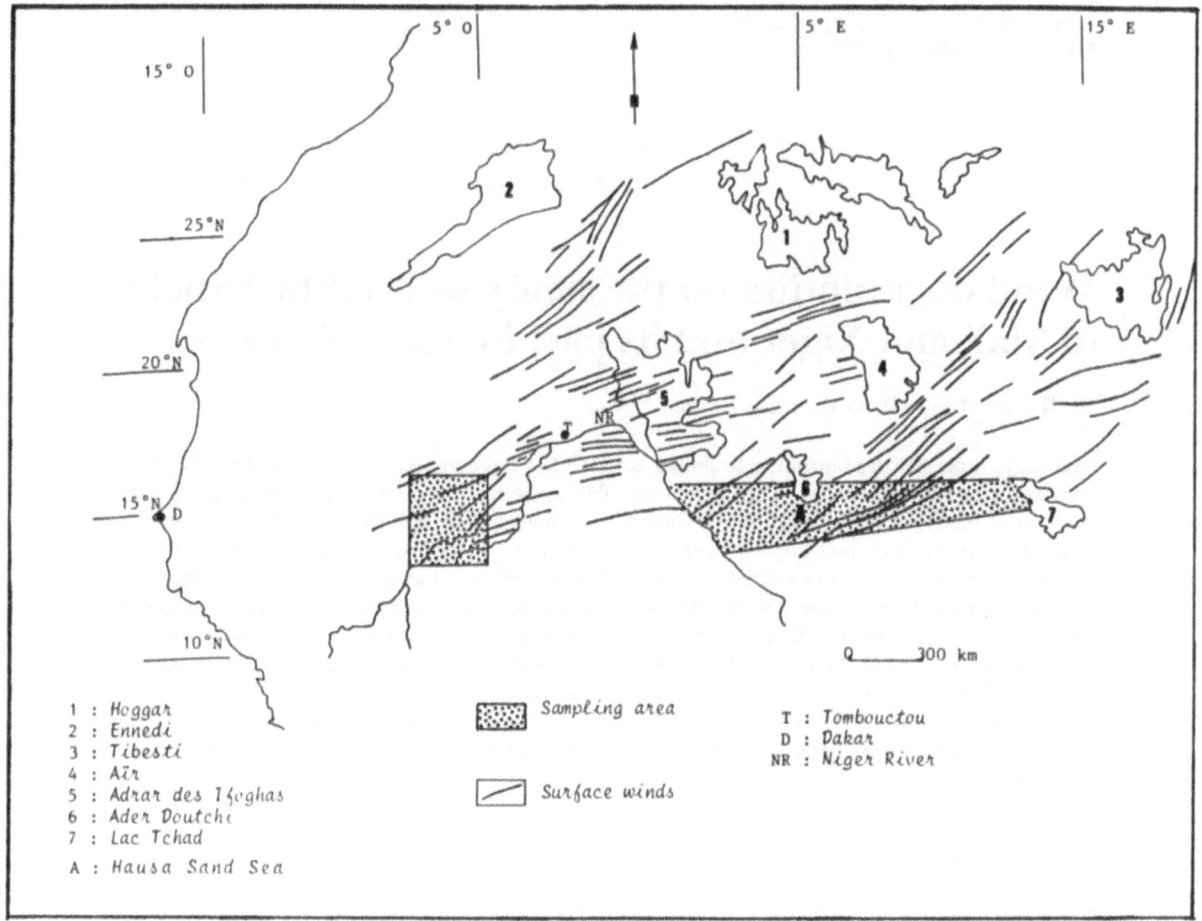

Fig. 1. Location of the two studied areas in the vegetated sand sea (Niger and Mali)

and ending with the semi-fixed sand ridges, the Goz in Sudan (Grove and Warren [5]). The southernmost latitude of stabilized dunes has been described in NW Nigeria in the Rima Sokoto river basin at about 12° 30 N (Sombroek and Zooneveld [26]).

The two areas, discussed in this paper, are located in Mali and Niger between 13° and 16° N, west and east of the river Niger bend (Fig. 1). They are in the centre of the Saharo-Sahelian sand belt leeward of the Tenere active sand sea where they form the fixed Hausa sand sea, a 600 km long sand accumulation extending from N'Guigmi (14° N—13° E) to Niamey (14° N—2° E).

The sand sea consists of different types of dunes (Fig. 2). We can consider that it begins at the latitude of the 150 mm isohyet with an area of transversal chains which, west of the Koutous massif, change to a hummocky and hilly dune ara. These sandy hills reach heights of 5—10 m and diameters of 60 to 400 m. They correspond to an *akle*[1] type dune field. They are the specific area of millet cultivation because of the favourable water balance of the soil in consequence of a favourable grain-size distribution and gentle slopes.

The analysed area in Mali is located leeward of Niger's inland delta in the region of Sokolo and Segou between 12° 30 to 15° N and 6° to 7° W. The stabilized sand sheet is

---

[1] An akle is a sand sea the topography of which is shaped in hummocks or hills preferentially aligned at right angle to the main wind direction.

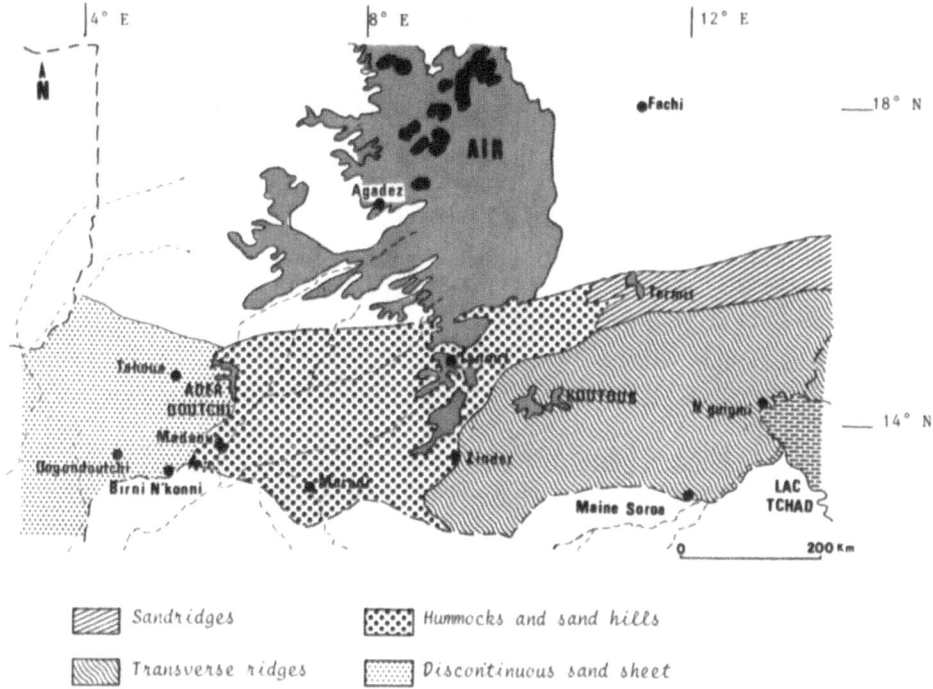

Fig. 2. Aeolian morphology in the Hausa sand sea

formed in the direction of the dominant wind, from east to west, in an area of transversal chains, gradually changing downwind to an area of longitudinal dunes. The top of these is eroded and their direction is between 65 and 70° ENE—WSW. Finally, more downwind, the dunes change to an area of hummocks.

Present climatic conditions interferring with paleoclimatic conditions determine the distribution and nature of the deposits in these drylands. There is evidence from past times that rainfall variations and droughts are the rule in African climates, particularly at the latitude of the present arid, semi-arid and sub-humid ecosystems (Nicholson [16]).

Climatic variations earlier than 20,000 BP are difficult to recognize because the indicators are rare and complicated to interpret. Scientists agree on three arid phases: 70,000 years BP, 20,000 to 15,000 years BP, and the last drought which began around 5,000 to 3,500 years BP. It is generally accepted that the peak of the last glacial period 18,000 BP corresponded to drought in tropical Africa (Sarnthein et al. [24]).

During this dry phase between 18,000 and 8,000 BP the southern limit of the Sahara (corresponding to the 150 mm isohyet) probably moved about 1000 km towards the north and between 6,000 BP and the present about 600 km towards the south (Petit-Maire [22]). This assertion is justified by the presence of an aeolian sand sheet at 5° further south than the present active dunes. This sand sheet can be followed from Senegal to Mali, Nigeria, Chad and Sudan. Its maximum depth of 60 m was observed in Niger (Mainguet and Chemin [12]), its maximum extension was probably Maiduguri in Nigeria (great erg of Hausaland).

Hurault [7] demonstrates that this dry climatic phase expanded until the actual Guinean ecozone (e.g. Banyo 6° 03 N). Hurault mentions another drought in the same area at the beginning of the Christian era.

Nicholson and Flohn [18], Street-Perrot and Roberts [27] show that after the glacial maximum there was increasing aridity at tropical latitudes culminating between 14,000 and 12,500 BP. Later, between 12,500 and 5,000 BP, the climate became progressively more humid with a belt of expanding lakes in Africa and in Arabia (Schneider [25] and Maley [13—14]). In the early Holocene (9,000 BP) tropical Africa was humid and lake Chad had a water surface of 320,000 km², more than twenty times the present surface.

Based on pollen and phytogeographic evidences Lezine [9] has shown a southward shift at 18,000 BP of the Sahelian wooded grassland to 10° N during the arid conditions. During a second northward movement of humid vegetation at 8,000 BP Guinean elements reach 16° N and Sahelian-Sudanian elements extended to the southern margin of the present Sahara (21° N) when the Atlantic monsoon flux increased. During the more humid phase between 12,500 and 5,000 BP two droughts could be identified around 10,200 BP and 7,400 BP. After 5,000—4,000 BP the rainfall decreased leading to a lowering of the lake levels (Lezine [9]).

In a study of the middle and late Holecene on the basis of a north-south transect from the Gilf Kebir at 24° N (Sudan) in the centre of the presently hyperarid eastern Sahara to the Wadi Howar area at 17° N, Pachur and Kröpelin [20] have shown that aridification affected the northern parts at 5,000 years BP and the southern parts at about 3,500 years BP (Haynes and Mead [6]). The northern fringe of the paleo-Sahel moved southwards approximately 500 km. Neolithic man seems to have intensified the effects of aridification.

The previous results show the rapid and numerous climatic variations in the zone which corresponds to the actual Sahel south of the Sahara: globally a dry phase from 20,000 BP to 12,500 BP, then a more humid phase from 12,500 to 5,000 BP, before reinstallation of dryness. This results in migration of the isohyets.

In dry periods the stabilized dune belt falls under a Saharan arid regime with sand accumulation in connection with a *positive aeolian sediment balance*, a sparse vegetative cover and soils exposed to erosion by wind and water. In this process wind erosion is always present while water erosion is more intermittent although important. The result of wind erosion is particle sorting in the framework of a unique *Wind Action System*. The coarser particles tend to be left behind, stones and gravels as *reg* in stony deserts. Coarse sand particles (Fig. 3) with a mode of 630 μm are found in the western Tenere. From here the particles become finer and finer in the wind direction, and mobile transversal chains which are indicator of a *positive sediment balance* are formed (Mainguet and Chemin [12]). Dune fields of rather finer particles tend to accumulate towards the more humid Sahelian boundary of the sub-Saharan zone where the sand becomes stabilized by a continuous cover of vegetation and the crests of the dunes are blown away. The fixation of these dunes is reinforced when the weather conditions turn more humid e.g. the Great Erg of Hausaland which lies 600 km south of the nearest active dune fields (Grove [4]).

Under hyperarid and arid conditions sand is exported to the margin of the savanna belt where it becomes fixed by the continuous vegetative cover. The boundary of the active dunes moves towards the south. Sand structures with round crests are replaced by sharp crests. If more humid conditions return, the boundary of the sand fixed by a vegetative cover moves to the north further towards the desert.

However, it must be emphasized that the immobilization of dunes by vegetation depends on the maintenance of this vegetative cover. Overcultivation or overgrazing, particularly during times of drought, destroy the binding of the sand by plant roots, expose the sand once more to deflation by wind and the material begins to move again. In addition wind blown sand can induce the erosion of clays and silts which otherwise would not be affected (McTainsh [15]). This is one of the origins of the fine suspensions of dust.

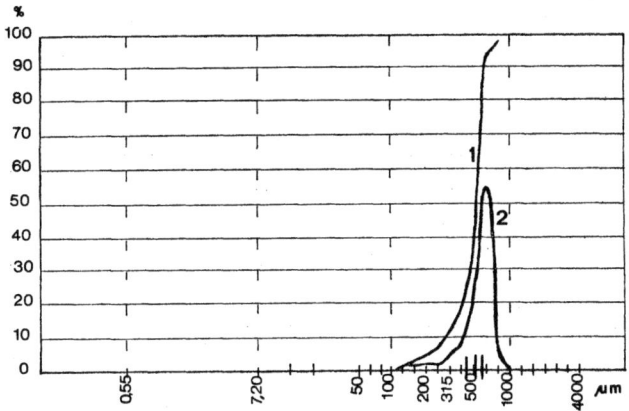

1 : Cumulative curve
2 : Frequency curve

Fig. 3. Distribution of the particles of a sandridge in the Tenere sand sea

1.2 The climatic and soil conditions

The Sahelian dry climate has in average a gradient of annual precipitation from north to south (Table 1).

Near Maradi (13° 30 N) the mean annual rainfall between 1932 and 1954 was 620 mm but only 430 mm between 1968 and 1975. These values show the significant variations in time of rainfall.

In general the old sand sea has a red ferruginous soil with iron and argilic beds in the highest part of the deposits. Along the dry wadis leaching is dominant, and the sandy soils are white and colorless. This type of soil evolution does not correspond to the actual annual rainfall between 400 and 500 mm in the area, but to the Sudano-Guinean ecozone with an annual rainfall of 1200 mm. (In Ivory Coast, the area of Tiebissou-Yamoussoukro-Bouafle.) The morphology of the soil profile in the Hausa sand sea retains the trace of more humid paleoclimates.

Coarse and fine sand fractions amount to over 90% and the pH varies from 5.0 to 9.0. The sandy soils are poor in nutrients and their humus content is low, but they are very permeable. In areas with large content of fine sand the water balance is excellent with relatively much available water during the dry season. Thus millet cultivation is favoured.

Table 1. Gradient of annual precipitation in the Sahel of Niger

16° N	Irregular number of rainfall days	150 mm
15° N	Less than 30 days of rainfall	300 mm
14° N	35 days of rainfall	550 mm
13° N	45 days of rainfall	700 mm
12° N	50 days of rainfall	830 mm

2 The aeolian dynamics

2.1 The wind currents

The two areas are located on the southern rim of the Saharo-Sahelian *Wind Action System*. Here the Harmattan (or continental trade-winds) blows from directions between NE—SW and ENE—WSW. The Nigerian region is swept by the most southerly NE—SW to ENE—WSW wind direction blowing around the Tibesti massif along two branches meeting on the windward side of the Bilma sand sea.

The Malian area is affected by two wind currents, one blowing in NE—SW direction north of the Hoggar massif and north of the Adrar des Ifoghas. The second one blows south of the Hoggar through the Adrar des Ifoghas corresponding to an ENE—WSW direction.

The main wind directions at the synoptic meteorological stations of Niger are shown (Fig. 4). Their frequency in the dry and humid seasons is proportional to the length of the arrows. The wind blows from NE to SW and becomes ENE to WSW in the Sahel (the Zinder region).

In Mali the mean annual wind rose (1982 to 1984) for the station of Nara (Fig. 5) shows that northerly and northeasterly wind directions are dominant with a maximum from the northeast for wind speeds of 4—7 m/s. These winds are the Harmattan winds which blow from November to February. From the southwest the monsoon winds blow from June to September.

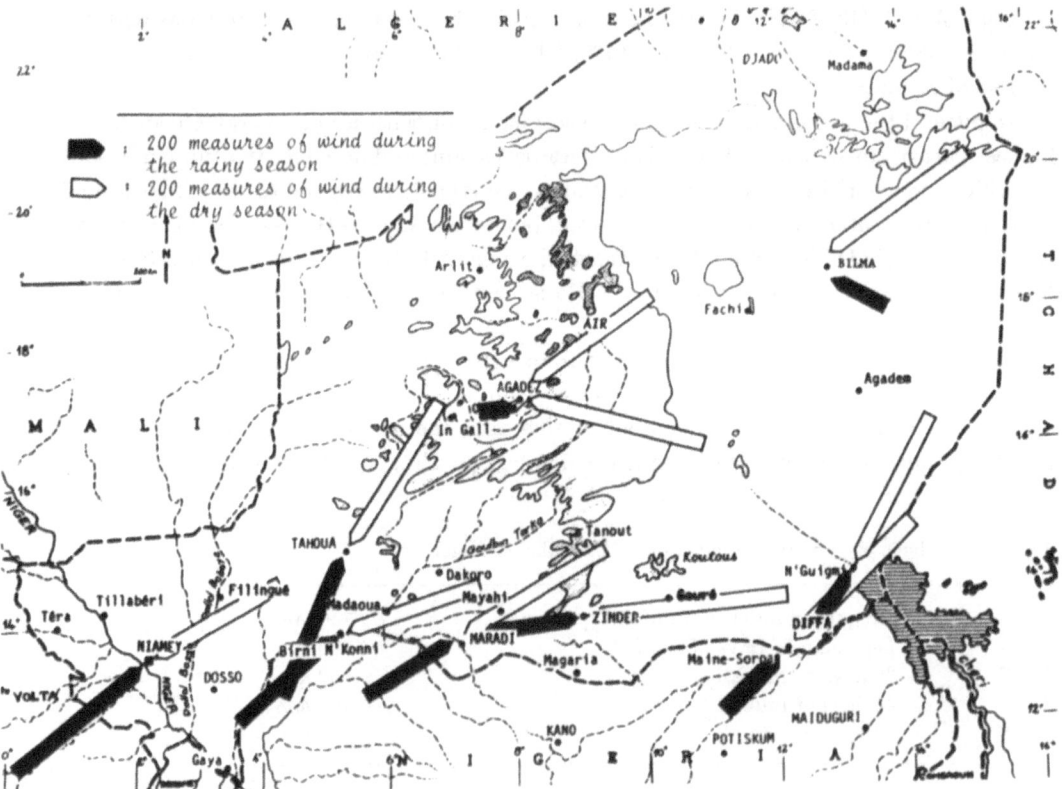

Fig. 4. Dominant wind directions in the synoptic stations of Niger, 1966—1975. (Each measurement represents one wind according to its direction without reference to its speed)

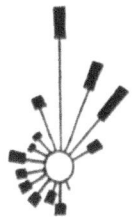

■■ *Five observations of wind velocity between 4 and 7 m/s*
— *Five observations of wind velocity < 4 m/s*

Fig. 5. Mean annual wind rose in Nara (Mali). From 1982 to 1984

2.2 Sedimentological aspects of the aeolian deposits

2.2.1 Sediment size distribution

Along a distance of 600 km the analysis of the sediment size distribution of the stabilized dune belt was based on 150 samples collected between N'Guigmi and Niamey in Niger and 40 samples from north, west and south of Segou (Mali) in the areas where deflation was maximum.

The sampling was made according to the classic geomorphological method: along several transects, samples were collected in each land unit as often as necessary when the land units were changing. Nevertheless on the fixed and therefore convex dunes, between N'Guigmi and Niamey, the samples were taken as often as possible at the top of the edifices. The granulometric analyses were made with a simple column of sieves from 50 to 2 000 μm with a progression of $2\sqrt{2}$.

To analyse the evolution of the size and the distribution in space the *mode* was calculated. A line joining the locations of samples with the same mode is called an isomode (Fig. 6).

— The smallest value, 80 μm mode, was found in a sample collected south of Segou (Mali). The deposits are allochtonous winnowed during the dry season from fluvial deposits in the river Niger, which is upwind.

— The 160 μm isomode is specific to Mali in the area of Sokolo and in the stations south of Niger from N'Guigmi to Niamey through Dosso, Dogondoutchi, Birni N'Konni, Madaoua, Maradi, Zinder. This is the most favourable material for millet cultivation. This mode corresponds also to the optimum of actual aeolian transport (Mainguet, Canon, and Chemin [10]). In Fig. 6 the 160 μm isomode has a curved shape reflecting the higher degree of degradation in the centre of the Hausa sand sea. The effect of deflation is maximum in the Maradi region where the intensity of land use is maximum.

— The 315 μm isomode is specific to the northern rim of the Malian area (Nara, Sokolo) and the Hausa sand sea (Tahoua, Tanout, Termit) in Niger. In the field these materials represent the deflation residue left by the winnowing effect of the Harmattan. The finest part of the sandy material has been exported by deflation.

— The 800 μm isomode is specific to the surface of the Tenere north of the Hausa sand sea where the whole system of sand ridges of the sand sea is mechanically blocked by "wind stable desert pavement" (Peel [21]). It can be found also in the cultivated area of Zinder where the aeolian degradation is maximum.

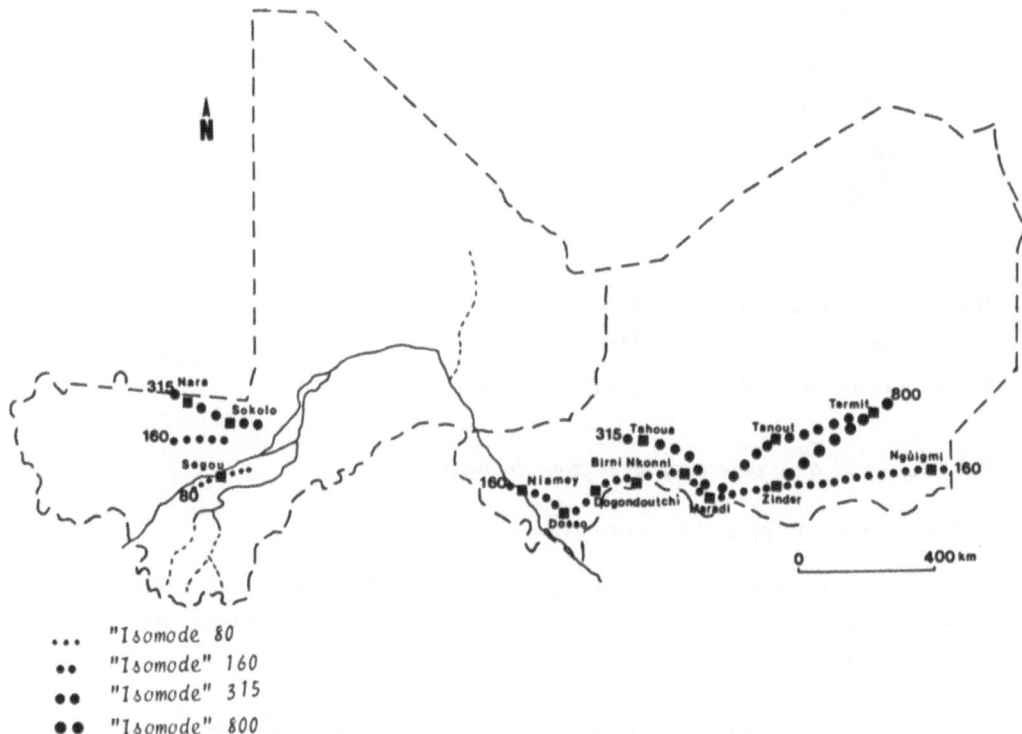

Fig. 6. Distribution and isomodes of the sands in Mali and Niger. The sliding of the curve of "315 microns isomode" towards the south in the area of Maradi results from deflation due to overuse of the sandy soils

The classification of the sands according to their size distribution (Fig. 7) reveals for all the samples four main types: fine particles easily reworked by deflation, medium sands divided in two classes, and coarse particles forming residual lag deposits.

— i: fine particles with a mode smaller than 125 μm are located in the areas of Segou (Mali) on the right bank of the Niger river.

— ii: medium sand particles with a mode between 125 and 250 μm which are dominant and accumulated in the areas of Sonango (Mali), Niamey, Dosso, Dogondoutchi, Birni N'Konni, Tahoua, Tanout, and N'Guigmi (Niger).

— iii: medium sand with a mode between 250 and 400 μm is the most abundant in the region between Maradi and Zinder (Niger) and Nara (Mali).

— iv: particles with a mode coarser than 400 μm form the highest proportion in the regions of Sokolo (Mali) and Termit (Niger).

These are the ranges found on reactivated previously vegetated and fixed sand dunes.

The regular and general distribution of the material along the aeolian transect for a distance as long as 600 km reveals its aeolian origin. The best sorting of the particles results from aeolian transport and accumulation or aeolian sorting processes. The width of the distribution has been quantified by the $Qd\Phi$ of Krumbein [8]. The $Qd\Phi$ is defined by

$$Qd\Phi = \sqrt{Q_1/Q_3},$$

where $Q1$ (25%) and $Q3$ (75%) are the quartiles. The scale of Φ is based on the logarithm with base 2 of the diameter.

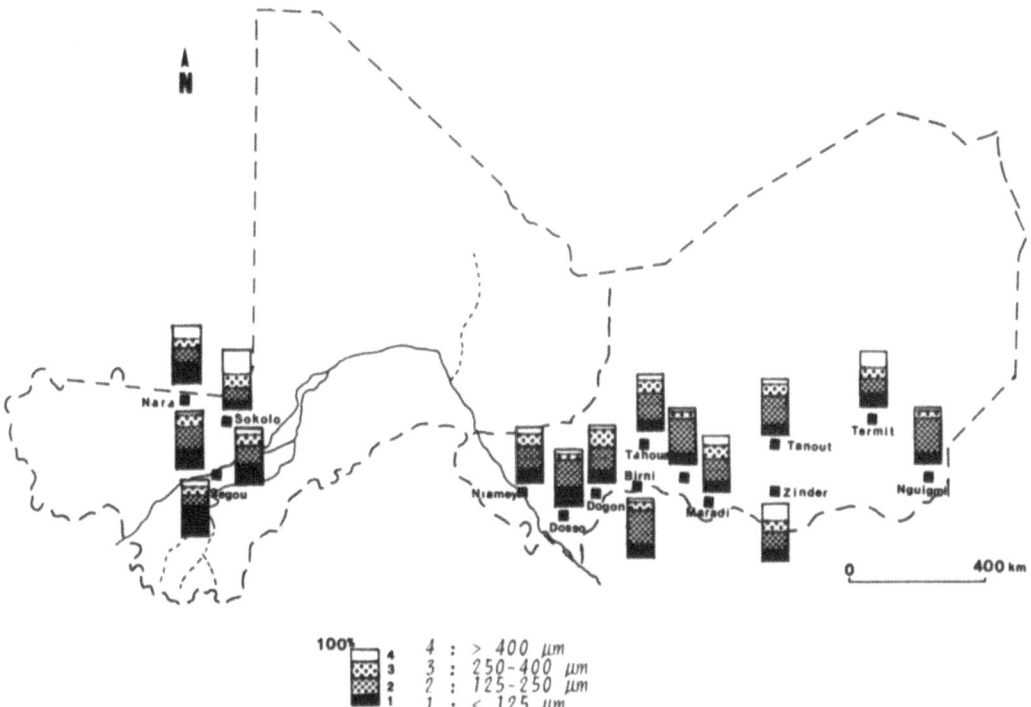

Fig. 7. Decrease of the sand size from east to west. The general trend is an increasing proportion of fine particles in the wind direction from ENE to WSW. The high proportion of coarse particles in the area of Zinder results from man induced deflation

The $Qd\Phi$ values are 0.53 for the sands with a mode of 160 µm near Segou (Mali), which is high. On the average the values are 0.15 to 0.24 for the sands between 160 and 315 µm in the Hausa sand sea.

The aeolian sorting was quantified by the ratio:

$$\frac{Qd\Phi \cdot 100}{M}$$

where $Qd\Phi$ is the Krumbein index and M the median. The ratio annuls the differences resulting from changes in the modal size and gives a homogeneous value of the aeolian sorting for the sands of Mali and Niger. The ratio increases from 0.03 at Termit (200 km northeast of the Hausa sand sea) to 0.45 at Segou (south of Mali).

As a conclusion a difference appears between particles smaller than 400 µm and those which are coarser. The smaller ones are reworked by the actual winds, the coarser are less affected by deflation.

2.2.2 Morphometric analysis

The shape of the sand particles can be quantified by the *abrasion index* (Fig. 8). The abrasion index of Cailleux [2] is derived from the Wentworth [28] roundness ratio and obtained by attributing a coefficient from 1 to 6 to grains according to their increasing roundness. The abrasion index was analysed (with a binocular) on 100 particles of each size class of each sample.

The value of the abrasion index analysed on the size classes 160, 315 and 500 µm shows

Fig. 8. Average value of the abrasion index measured and calculated on the sizes of 160 μm, 315 μm, and 500 μm. The abrasion index is high in the eastern area for the coarse particles (500 μm), and high in the western area for the medium particles

a constant decrease in the Harmattan direction. The lowest values (129—140) were observed in the south of Mali, south of the Niger river and in the western Niger in the region of the Dallols (dry valleys). They all correspond to fluvial autochtonous material briefly reworked by wind. The highest values (180—210) were observed in the regions of Nara and Sokolo in Mali and in the eastern part of the Hausa sand sea in Niger, where they correspond to the active crests of reactivated dunes.

The combination of the grain size-distribution and the abrasion index shows, at the scale of the synoptic wind action system the constant decrease of the material size and simultaneously the increase of wind abrasion on the particles. This demonstrates the existence of a unique unit and the regular effect of wind along wind trajectories. Discontinuities in this general east-west trend exist when fluvial accumulations, brought by rivers streaming southwards, like the Dallols or Niger, are accumulated. Then the granulometric gradient becomes more irregular.

3 Wind erosion, land degradation or desertification

3.1 Qualitative approach to deflation

According to their size sand particles have different behaviours. Particles smaller than 80 μm are transported in suspension by the harmattan. During the last three decades (1957—1987) the duration of dust storms have increased remarkably in Niamey as shown in Fig. 9 (Bertrand and Legrand [1]). Thus, the trend of the Sahel to become a source area for dust is confirmed.

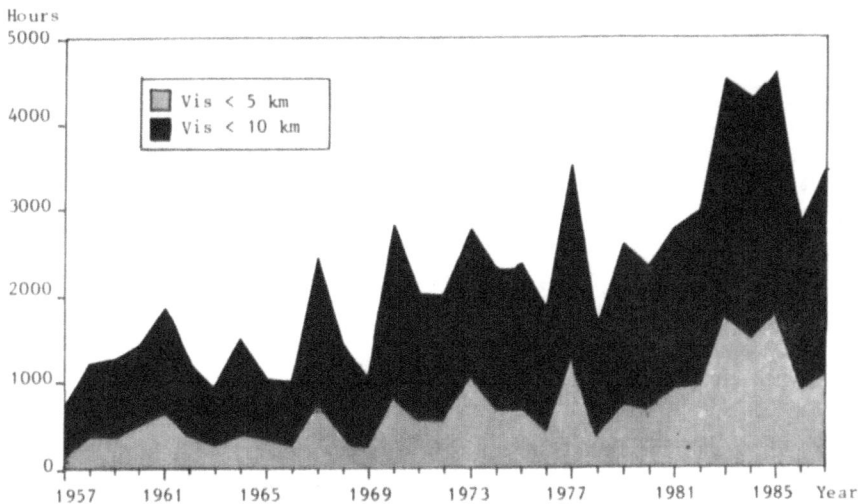

Fig. 9. Niamey (Niger): Time serie of the number of hours with reduced visibility at the soil surface to less than 5 and 10 kilometers. (After [1]).

Dust storms, blowing over the Hausa sand sea, often disturb aircraft and have their origin more to the northeast in two main locations in the Bilma and the Termit regions. Dust storms mainly occur during the months from October to May. Dust plumes pass over West Africa and cross occasionally the Atlantic Ocean (Prospero et al. [23]).

In these areas saline efflorescences, favouring aggregate formation, increase the drag of the wind. The minimum wind speed required for transport can move grains of 100 μm. Coarser particles require greater wind speed.

In addition to the particle supply of the "Harmattan dust aeolian system" (McTainsh [15]) dust comes also from deflation of the destabilized Hausa dune belt.

Fine and medium sand particles between 80 and 400 μm can be reworked by saltation and reorganized in sand dunes of the type of seifs (linear dunes) when the topography is irregular and the wind current divided in two branches. Barchanic dunes are formed where the topography is more regular and the wind regime unidirectional.

Coarser sand particles, larger than 400 μm, are not transported, but are locally moved by creeping. They are the residual part of the winnowing process and lead to sterile pavements or coarse ripples.

The fine sandy area where the sand sheet is composed mainly of particles between 80 and 160 μm (regions of Sonango and Segou in Mali and the western part of the Hausa sand sea) is submitted to severe aeolian deflation. Potentially the modal dimension corresponds to the grain size most likely to be lifted by aeolian forces. Field observations show severe deflation, impoverishment in organic matter, appearance of nebkas (sand arrows leeward of small obstacles: stones, bushes). In the area of Segou where the sand sheet gets thinner because of its location almost at the southern fringe of the Sahelian stabilized dune belt, the dangerous consequence of deflation is a total disappearance of cultivable land and exposure of iron pans or basement rocks.

The middle sandy areas of Nara (Mali) and Maradi, Zinder (Niger) are submitted to very severe deflation with loss of cultivable soil, loss of fertility, decreased yields of millet with only 200–250 kg/ha. Thus the farmers have to cultivate larger and larger areas, still farther from their villages, and they have to reduce the length of the fallow periods.

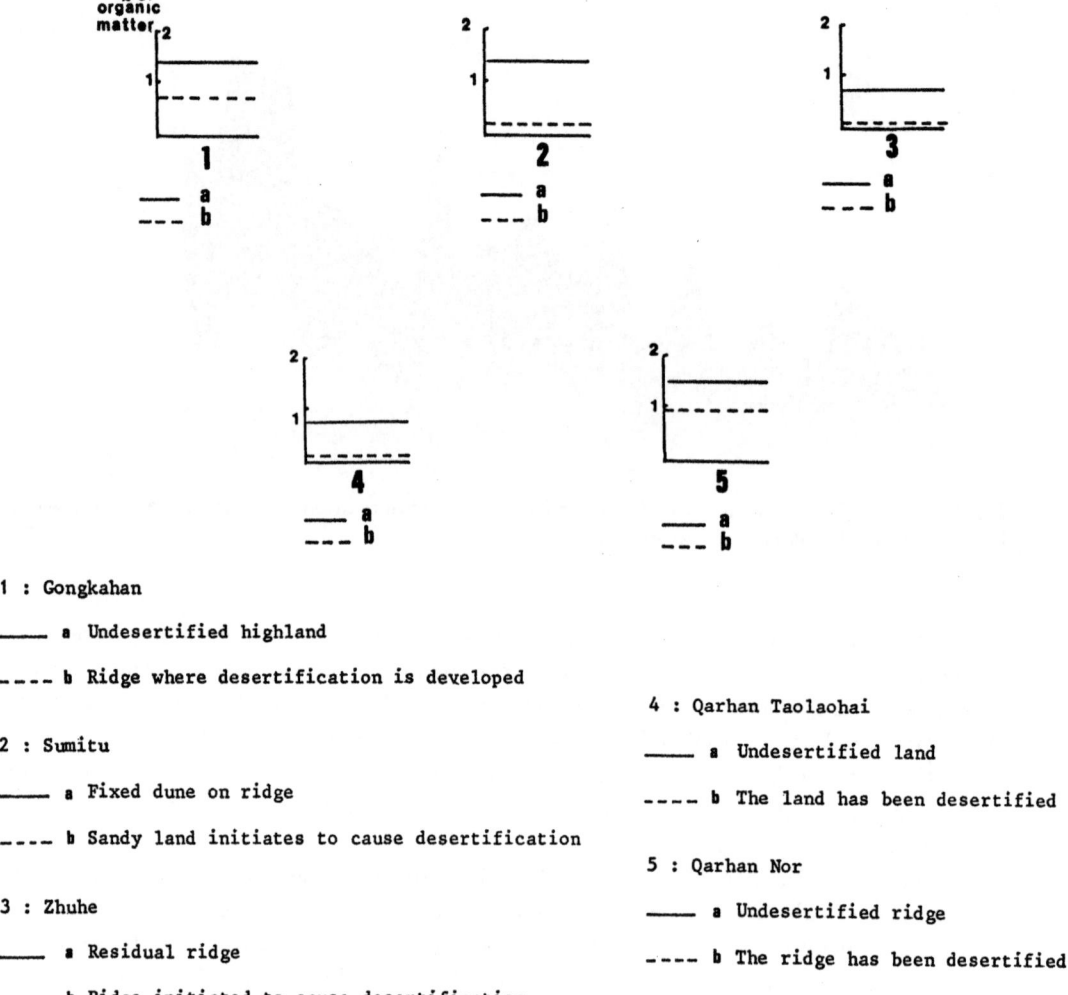

1 : Gongkahan

_____ a Undesertified highland

____ b Ridge where desertification is developed

2 : Sumitu

_____ a Fixed dune on ridge

____ b Sandy land initiates to cause desertification

3 : Zhuhe

_____ a Residual ridge

____ b Ridge initiated to cause desertification

4 : Qarhan Taolaohai

_____ a Undesertified land

____ b The land has been desertified

5 : Qarhan Nor

_____ a Undesertified ridge

____ b The ridge has been desertified

Fig. 10. Loss of organic matter in the topsoil of cultivated soils in Ordos Steppe (China). (Measures by Zhu Zhenda and Liu Shu [29])

The measurements confirm that 400 μm is the upper limit for actual aeolian deflation since the smaller particles are absent from the surface samples of deflated areas.

Areas with coarse sand pavement with modes between 630 and 800 μm seal the surface to wind erosion and are unfavourable for agriculture. Deflation is maximal in Niger in the region between Termit and the eastern slope of the Ader Doutchi massif. Leeward of the river Niger bend in Mali deflation has reached its maximum strength in the northern two thirds of the sand cover.

3.2 Quantitative evaluation of deflation

Quantitative analysis of the size distribution of the residual topsoil and the subsoil can be used for an evaluation of the effects of deflation.

The effects of wind erosion on the top soil in sandy soils, analysed on similar types of

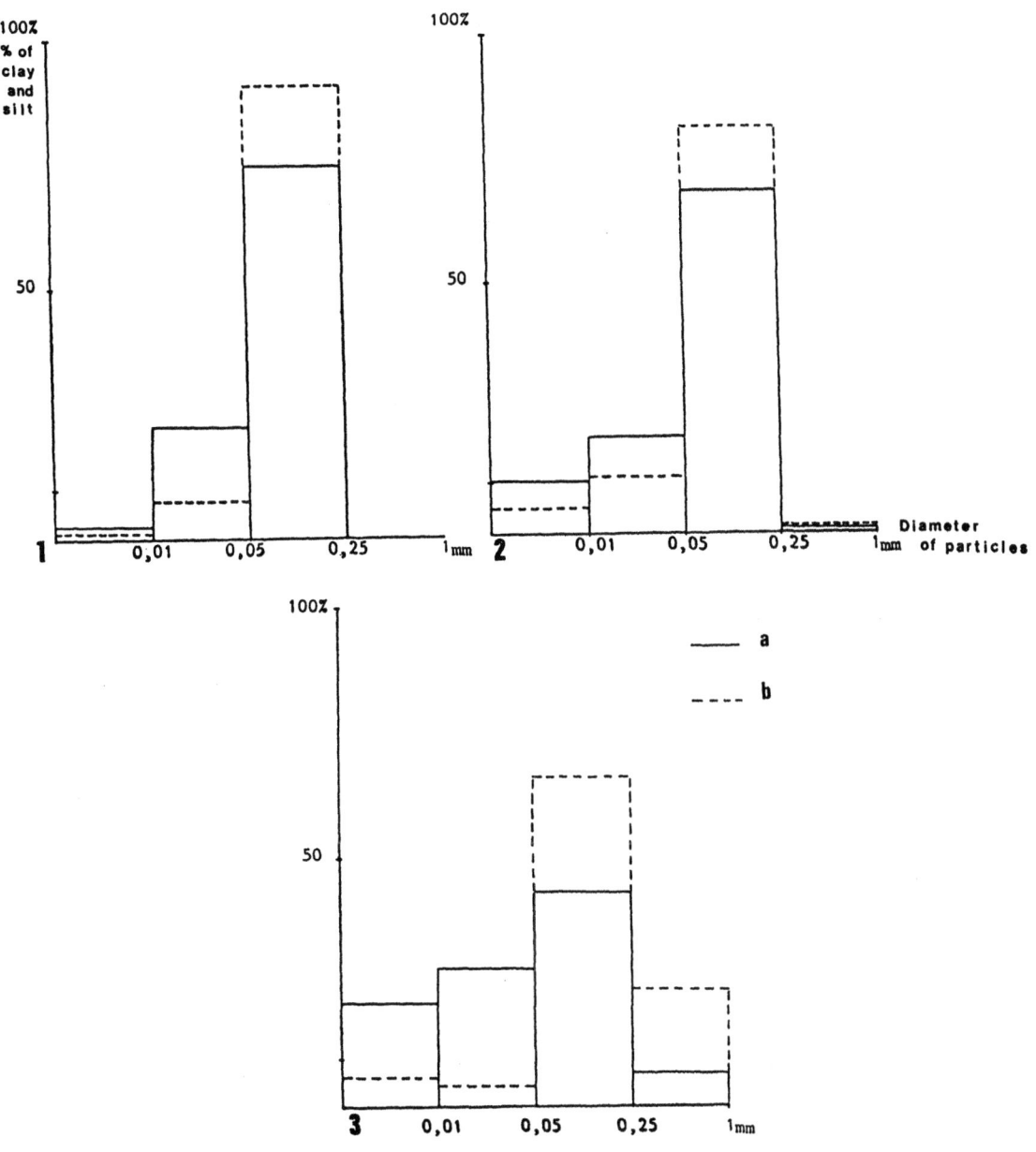

Fig. 11. Loss of clay and silt in cultivated soils. Increase of the percentage of sand in the top soils as a result of reactivation of vegetated dunes in Ordos Steppe (China). (Measured by Zhu Zhenda and Liu Shu [29])

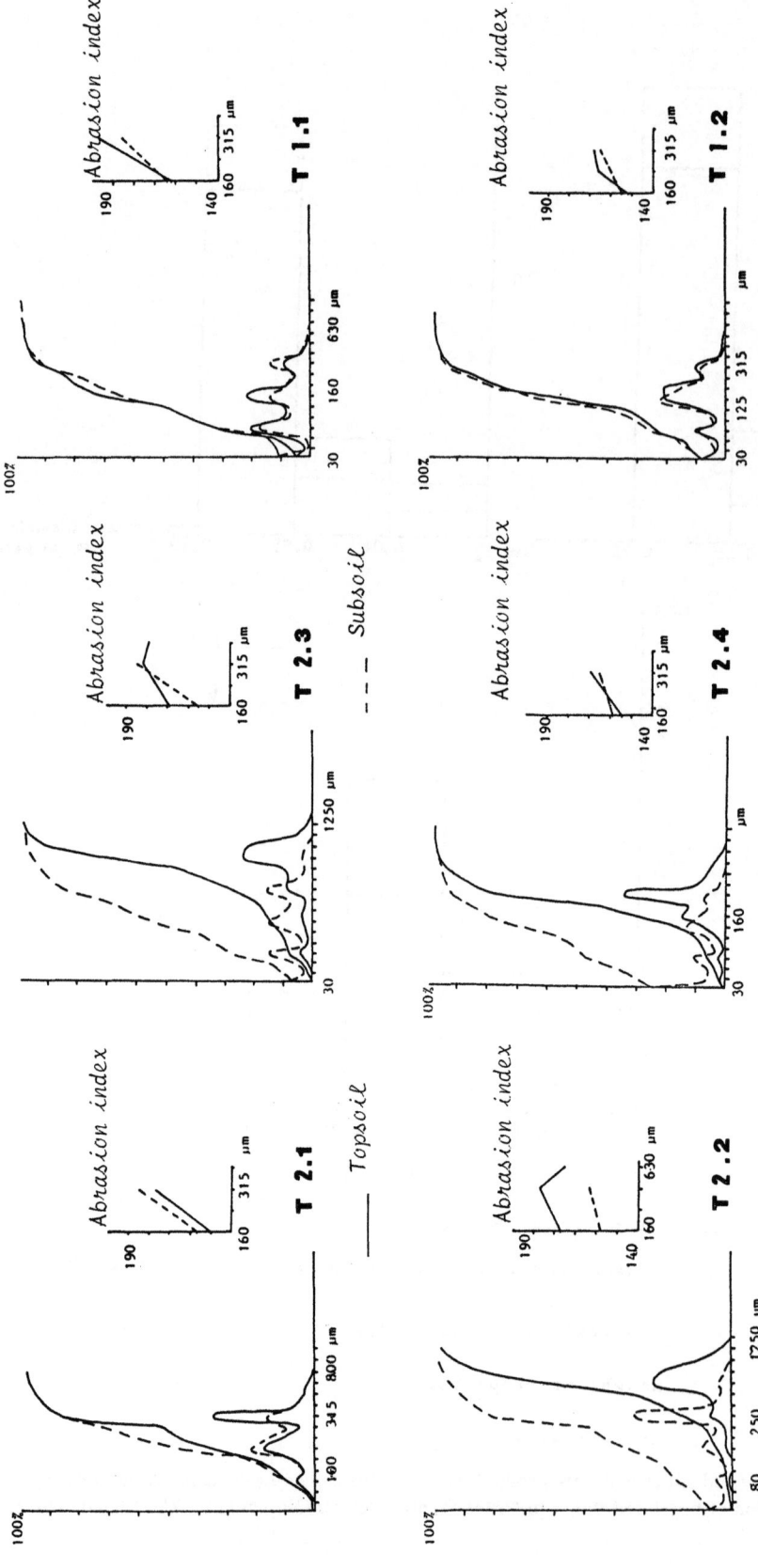

Fig. 12. Comparison of the size distribution in samples collected at the same location (Niger) in the topsoil — and the subsoil ---

Table 2. Types of soil degradation by wind effects

Examples	Topsoil	Subsoil	Threshold of lift up (deflation)
Case 1:			
T. 1.1. (Fig. 12) Sandridges 14° 55 N	On residual plots Mode: 160 microns 75% < 200 microns	Outcropping Mode: 80 microns 75% < 250 microns	
T. 1.2. (Fig. 12) 13° 55 N	Bimodal: 160—200 microns 75% < 200 microns	Mode: 160 microns 75% < 200 microns	
Case 2:			
T. 2.1. (Fig. 12) Gullies 14° 55 N	With a coarse sand pavement Mode: 315 microns 50% > 250 microns	With a hard setting Mode: 160 microns 75% < 250 microns	
T. 2.2. (Fig. 12) Transversal Akle	Mode 630 microns 75% > 400 microns	Mode: 315 microns 75% < 250 microns 25% < 125 microns Hard setting	400 microns
T. 2.3. (Fig. 12) Sandridges 14° 35 N	Mode 630 microns 75% > 315 microns	Trimodal: 80—160—315 microns 75% < 250 microns	315 microns
T. 2.4. (Fig. 12) Ripple-marks 13° 50 N	Mode 315 microns 75% > 200 microns	Mode 30 microns 75% < 200 microns Hard setting	200 microns

vegetated dunes in the Ordos steppe by Zhu Zhenda and Liu Shu [29] bring a good comparison and show:

1) Removal of organic matter. Measurements as shown in Fig. 10 quantify the considerable loss of organic matter on all sites where the semi-arid vegetation has disappeared and the surface has been reworked into active sand structures.
2) Changes in the texture with increase of the coarse sand material of over 50 μm and a proportional decrease of the fine particles smaller than 50 μm (Fig. 11).
3) Changes in the porosity of the top soil from 7 to 1%.

In Niger two types of degradation of the top soil can be seen (Table 2):

Case 1: (T. 1.1.—1.2., Fig. 12). The top soil (A horizon) subsists only in residual patches; the sandy material is fine to medium, an optimal size for aeolian deflation. The differences between the top soil and the sub soil are very small, but with a slight increase in the proportion of fine material (clay and silt) in the sub soil (B horizon). These differences decrease in the wind direction from north and east to south and west.

Case 2: (T. 2.1.—2.2.—2.3.—2.4., Fig. 12). The cultivable A horizon is totally eroded by wind erosion, the coarser winnowed residues are forming some round spots or ripple marks because the size of the material is greater than the deflation capacity.

Under these residues the sub soil is sealed by *hard-setting* (or crusting) called *walla-walla* in the Hausa language. Here soil degradation is a result of external wind erosion and from intermediate degeneration of the structure leading to the formation of hard-setting.

To the negative aeolian sediment balance of case 1 an effect of water erosion is added in case 2. Crusting will only occur if the top soil has disappeared and the sediment balance has changed from positive to negative in connection with over-use and drought. Six analysed samples (Table 2) quantify the effect of soil degradation. It is shown that wind deflation affects the surface until a particle threshold of $400 \, \mu m$ has been reached in transversal sand structures with positive sediment balance. Moreover, until a particle size of $315 \, \mu m$ has been reached more down wind where the structures are becoming longitudinal sand ridges, and where the sediment balance is negative.

4 Conclusions

4.1 The deposit of a deep sand sheet at the latitude of the present Sahel can only be explained by a dryer paleoclimate during which the Sahel was in the southern part of the *Sahara Wind Action System*. It was during this dry climatic phase that an accumulation area with a *positive sediment balance* was formed.

4.2 In the wind direction, for a distance of 600 km, the size gradient and the distribution of aeolian sand particles demonstrate that the dynamics of sand accumulation is due to the same *Wind Action System*.

4.3 During the last discontinuous drought (1968—1983) the sediment balance became negative because of degeneration of the vegetative cover and overuse for grazing and farming (mainly for sorghum and millet cultivation). Therefore the Sahel has turned to a source area of dust from which particles are removed.

4.4 The *negative sediment balance* is not acting with the same severity in all the areas. The severity of the deflation can be classified according to the size index and the mode of the sandy soils. The most vulnerable areas are the most densely occupied by man and cattle, but this occupation is not accidental. It corresponds to an advantageous situation where the grain size distribution is favourable to a good water balance in the soil.

4.5 In the northern fringe of the stabilized dune belt sand deposits, which are coarse in consequence of deflation, result in a stable, coarse sand pavement resistant to wind erosion. To the south, where the material is finer, the removal will continue until a complete disappearance of the sand sheet and exposure of the iron pan or the hard rock.

4.6 The vulnerability of the Sahelian vegetated dune system depends more on the sensitivity of the mobilizable particles, favoured by the degradation of the vegetative cover' than on the wind velocities.

4.7 Dunes and sand accumulation are often mentally associated with strong wind and arid conditions. This interrelationship must be reviewed. Particle deposition requires mechanisms which decrease the wind speed and the particle transport. When aridity is increasing, wind velocity and wind frequency are increasing, and the trend is not aeolian accumulation, but wind erosion.

References

[1] Bertrand, J. J., Legrand, M.: Télédétection des brumes sèches. Colloque intern: la télédétection au service du développement (1987).

[2] Cailleux, A.: L'indice d'émoussé: définition et première application. C. R. Soc. Géol. Fr. 13—14, 251—252 (1947).

[3] Collective: Wind erosion conference. Lubbock, U.S.A. 1988.

[4] Grove, A. T.: The ancient erg of Hausaland and similar formations in the south side of the Sahara. Geogr. J. CXXIV-4, 528—533 (1958).

[5] Grove, A. T., Warren, A.: Quaternary landforms and climate on the south side of the Sahara. Geogr. J. 134, 194—208 (1968).

[6] Haynes, C. V., Mead, A. R.: Radiocarbon dating and paleoclimatic significance of subfossil Limicolaria in northwestern Sudan. Quaternary Res. 28, 86—89 (1987).

[7] Hurault, J.: Phases climatiques tropicales sèches à Banyo (Cameroun, hauts plateaux de l'Adamawa). In: E. M. Van Zinderen Bakker (ed.): Palaeoecology of Africa, the surrounding islands and Antarctica, pp. 93—102. Cape Town: Balkema (1972).

[8] Krumbein, W. C.: Size frequency distribution of sediments and the normal ϕ curve. J. Sediment. Petrol. 8—3, 84—90 (1938).

[9] Lezine, A. M.: Late quaternary vegetation and climate of the Sahel. Quaternary Res. 32, 317 to 334 (1989).

[10] Mainguet, M., Canon, L., Chemin, M. C.: Le Sahara: géomorphologie et paléogéomorphologie éoliennes. In: Williams, M. A. J., Faure, H., (eds): The Sahara and the Nile — quaternary environments and prehistoric occupation in northern Africa, pp. 17—35. Rotterdam: Balkema 1980.

[11] Mainguet, M.: Les marques géographiques de l'aridité dans le Sahara et sur ses marges au Plio-Quaternaire. A-t-on fait une juste place au Paléo-aride du Sahara. Bull. A. G. F. 483—484, 25—29 (1982).

[12] Mainguet, M., Chemin, M. C.: Sand seas of the Sahara and Sahel: an explanation of their thickness and sand dune type by the sand budget principle. In: Brookfield, M. E., Ahlbrandt, T. S., (eds.): Eolian sediments and processes, pp. 353—363. Amsterdam: Elsevier 1983.

[13] Maley, J.: Mécanismes des changements climatiques aux basses latitudes. Palaeogeography, Palaeoclimatology, Palaeoecology 14, 193—227 (1973).

[14] Maley, J.: Etudes palynologiques dans le bassin du Tchad et paléoclimatologie de l'Afrique nord-tropicale de 30,000 ans à l'époque actuelle. Travaux et documents de l'ORSTOM 129, p. 586 (1981).

[15] McTainsh, G.: Dust processes in Australia and West Africa: a comparison. Search 16-3-4, 104—106 (1985).

[16] Nicholson, S. E.: A climatic chronology for Africa: synthesis of geological, historical and meteorological information and data. Ph. D. Thesis, Wisconsin-Madison, unpublished 1976.

[17] Nicholson, S. E.: Climatic variations in the Sahel and of Africa regions during the past five centuries. J. Arid Environ. 1, 3—24 (1978).

[18] Nicholson, S. E., Flohn, H.: African environmental and climatic changes and the general atmospheric circulation in Late Pleistocene and Holocene. Climatic Change 2, 313—348 (1980).

[19] Nicholson, S. E.: The historical climatology of Africa. In: Wigley, T. M. I., Ingram, K. J., Farmer, G., (eds.): Climate and history, pp. 249—270. Cambridge: 1981.

[20] Pachur, H. J., Kropelin, S.: Wadi Howar: paleoclimatic evidence from an extinct river system in the south eastern Sahara. Science 237, 298—300 (1987).

[21] Peel, R.: Wind-stable stone-mantles in the southern Sahara. Geogr. J. 134, 463—465 (1968).

[22] Petit-Maire, N.: Interglacial environments in presently hyperarid Sahara: palaeoclimatic implications. Palaeoclimatology and Palaeometeorology: modern and past patterns of global atmospheric transport, 637—661 (1989).

[23] Prospero, J. M., Glaccum, R. A., Nees, R. T.: Atmospheric transport of soil dust from Africa to South America. Nature 289, 570—572 (1981).

[24] Sarnthein, M., Tetzlaff, G., Koopman, B., Walter, K., Pflaumann, V.: Glacial and interglacial wind regimes over the eastern subtropical Atlantic and north west Africa. Nature 293, 193—196 (1981).

[25] Schneider, J. L.: Evolution du dernier lacustre et peuplements préhistoriques aux pays bas du Tchad. Bull. Inst. Fr. Afr. Noire A31, 259—263 (1969).

[26] Sombroek, W. G., Zonneveld, I. S.: Ancient dunefield and fluviatile deposits in the Rima-Sokoto
     river basin (NW Nigeria). Soil Survey Papers 5, 102 (1971).
[27] Street-Perrott, F. A., Roberts, N.: Fluctuations in closed-basin lakes as an indicator of past atmo-
     spheric circulation patterns. In: Street-Perrott, Beran, Ratchliffe (eds.) Variations in the glo-
     bal budget, pp. 331—345. Dordrecht: Reidel (1983).
[28] Wentworth, C. K.: A scale of grade and class terms for clastic sediments. J. Geol. 30, 377—392
     (1922).
[29] Zhu Zhenda, Liu Shu: Combating desertification in arid and semi-arid zones in China. Institute
     of desert research Academia Sinica, p. 69 (1983).

Authors' address: M. Mainguet and M. C. Chemin, Laboratoire de Géographie Physique Zonale —
Université, 58 rue P. Taittinger, F-51100 Reims, France.

Acta Mechanica (1991) [Suppl] 2: 131—146

The effect of sea cliffs on inland encroachment of aeolian sand

H. Tsoar and **D. Blumberg**, Beer Sheva, Israel

Summary. The effect of sea cliffs on aeolian sand encroachment inland was monitored by means of traps measuring the sand transport and microanemometers measuring wind flow, because these are affected by the topography of the cliffs. Measurements were carried out on two cliffs in the southern coastal-plain of Israel: one 160 m and the second 240 m long, with front slope inclination of 30° to 40° and a height of 22—25 m. Results show that sand is incapable of climbing even moderate cliff slopes of 10°—15° because storm wind impinges on the cliffs at angles of 45° to 50° to the rim of the cliffs and is diverted to flow in a helical pattern parallel to the shoreline along the front slope of the cliffs. The diverted sand-moving wind is again deflected at the northern end of the cliffs, where its magnitude abates and it thus deposits sand that is being carried along the beach. The flow of the wind on the beach parallel to the foot of the cliff represents the effect of a non-homogeneous secondary wind. The wind is accelerated along the cliff, and as a result sand transport increases along the cliff. After a length of 150—200 m along the front slopes of the cliffs a state of equilibrium is achieved.

1 Introduction

1.1 Coastal dunes and their relationship to sea cliffs in the Southeastern Mediterranean

Sand dunes have encroached the southeastern Mediterranean coast, from the Bardawil Lagoon in Sinai up to Haifa Bay in Northern Israel (Fig. 1). North of Tel-Aviv, in the Sharon region, there is a long, steep, continuous sea cliff made of aeolianite with a maximum height of 50 m above a beach 20—30 m wide. Encroachment of aeolian sand along this part of the coast only occurs where there are breaches in the cliff. South of Tel-Aviv, in the southern coastal-plain of Israel, the beach is wider (up to 50 m) and sea cliffs are less common; when they occur they are low (a maximum height of 25 m) and discontinuous. The coastal dune strip in this area is up to 6 km wide.

1.2 Research hypotheses

The reason for the inland far-reaching encroachment of sand onto the southern coastal-plain and its nonoccurence in the Sharon region can only be because of the sea cliff. The postulate that sand encroachment in the southern coastal-plain takes place because the amount of sand reaching its beaches by longshore currents is double that reaching the Sharon beaches by the same currents (see e.g. [9] [13]) does not hold, because, where the sea cliff is nonexistent or breached in the Sharon region, sand does penetrate and wide coastal dune fields do develop.

The conjecture is that the formative characteristics of a sea cliff, such as height, length,

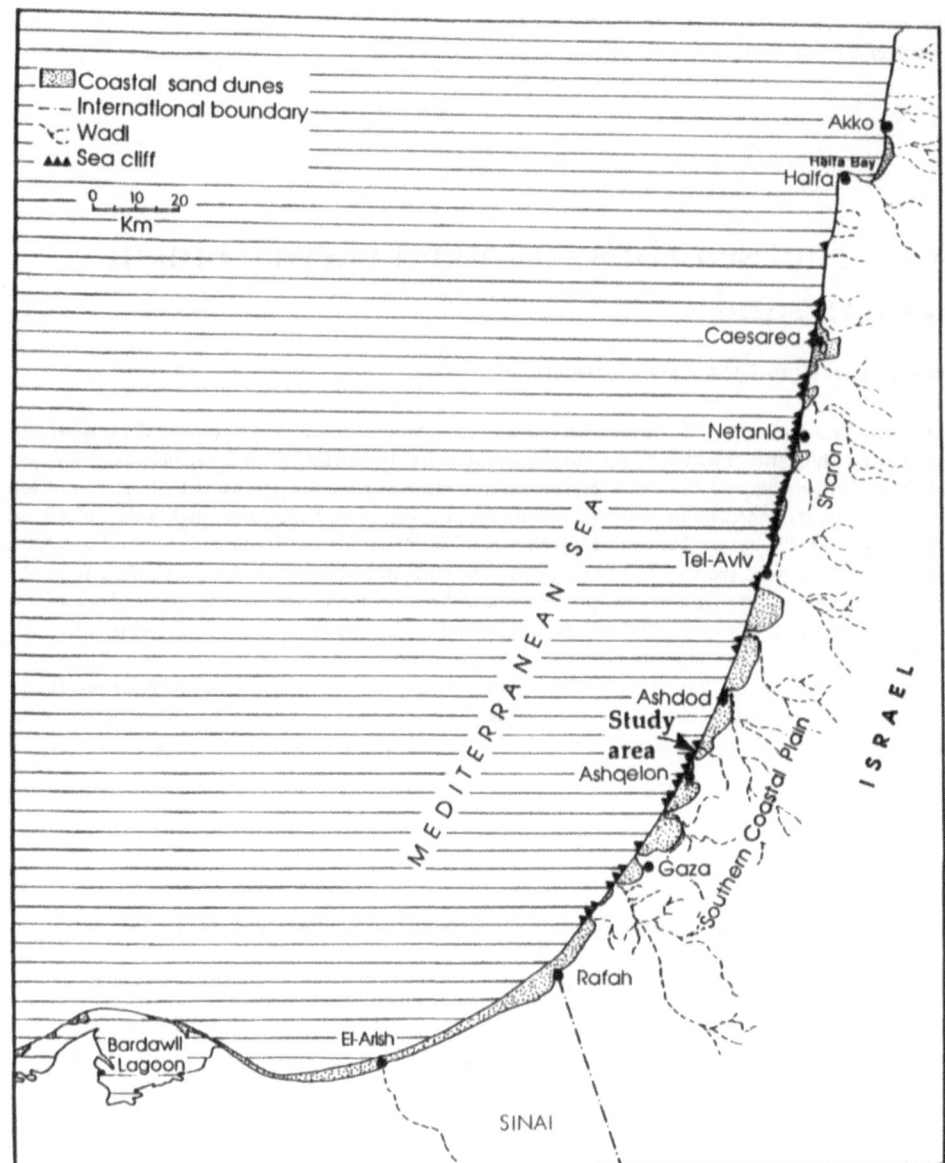

Fig. 1. Map of the coastal dunes of the southeastern Mediterranean coast. The research area is between Ashdod and Ashqelon. Sea cliffs are marked by black triangles (after Tsoar [17])

steepness, and also the width of the beach between the shoreline and the foot of the cliff, could be factors that prevent or facilitate encroachment of sand inland. The effect of these factors on aeolian sand encroachment were monitored in the field on two adjacent cliffs during storms.

1.3 Research area

The two cliffs under investigation consist of aeolianite and are located in the southern coastal-plain of Israel (Fig. 1). The northern cliff has a maximal height of 22 m and a length of 160 m, whereas the southern one is 25 m high with a length of 240 m. Between them there

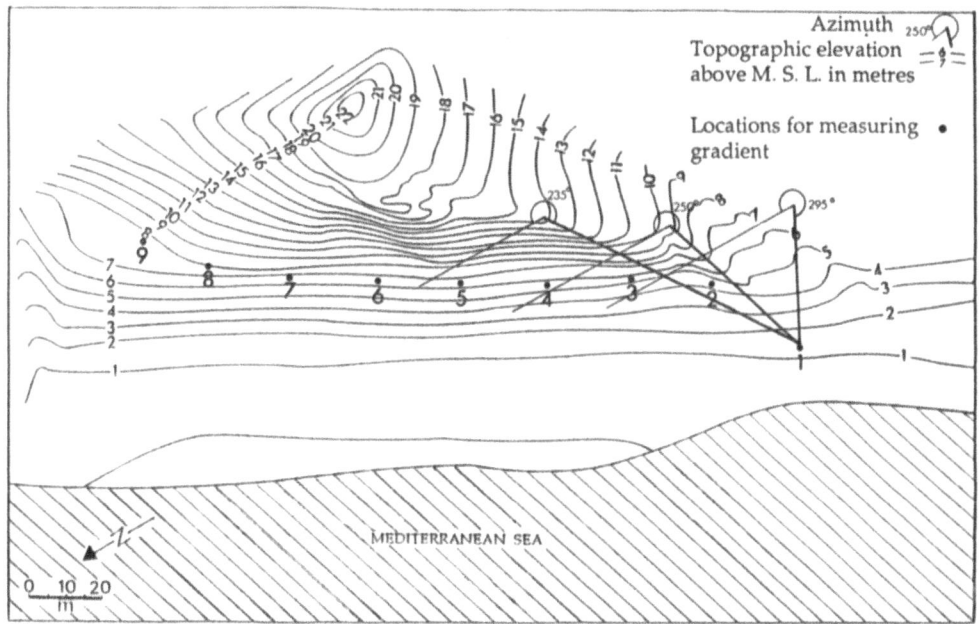

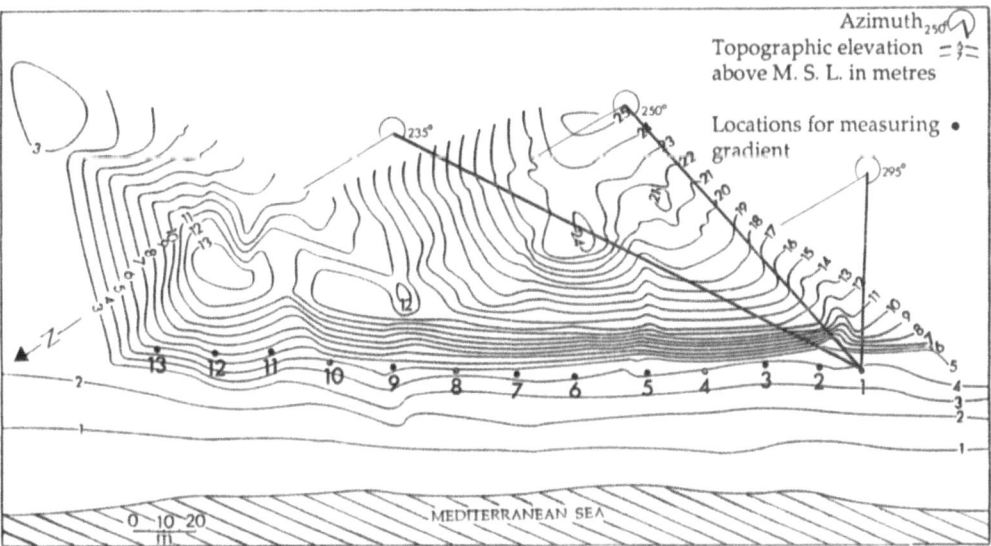

Fig. 2. Maps of the two cliffs of the research area and the various azimuths on which slope declinations were taken (Tables 1 and 2). **A** northern cliff, **B** southern cliff

is a 150 m long gap. A parabolic dune field, 3—3.5 km wide, runs east of the cliffs, which demonstrates that these cliffs do not represent an obstacle for aeolian sand encroachment.

The cliffs were surveyed by an electronic theodolite. Figure 2 depicts maps of the cliffs and Tables 1 and 2 show the cliff slope declinations along different azimuths at an average 12 m horizontal distance. The maximal declination perpendicular to the cliff is 40°. The maximal declination along a 0.5 m horizontal distance is 70°.

Figure 3 depicts maps of the two cliffs showing the source of the sand on the cliffs and their surroundings. There are vast aeolian sand deposits on the northern flanks of both cliffs, whereas their southern flanks are completely devoid of sand.

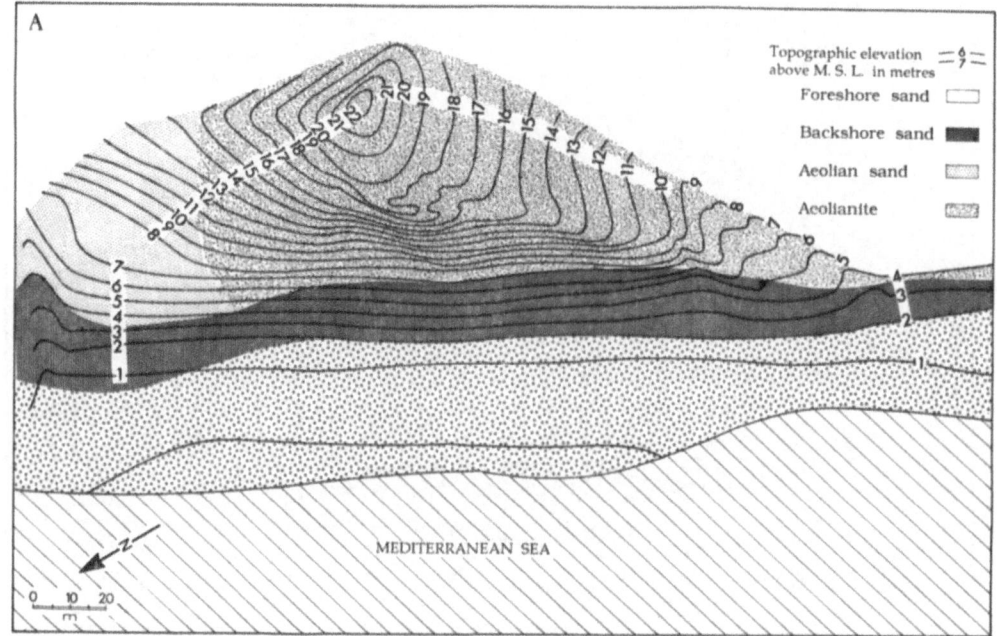

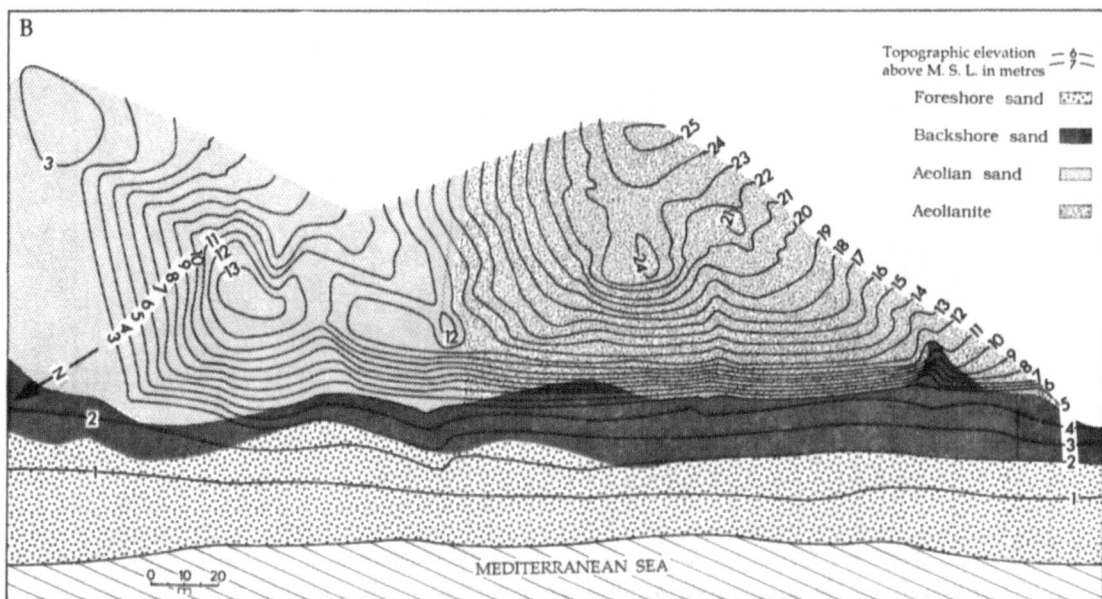

Fig. 3. Maps of the aeolianite cliffs and of the sand types of the various environments at the research area. **A** northern cliff, **B** southern cliff

1.4 Wind regime at the research area

The threshold velocity for the aeolian sand transport on the beach was estimated as 7 m s^{-1} (measured at a height of 40 m). Wind measurements taken at that height at both Ashdod and Ashqelon by the Israeli Meteorological Service indicate that, in winter 19.3% of the time the wind velocity was between 7 and 13.9 m s^{-1} and 2.7% of the time between 14 and 20.9 m s^{-1}. In summer the magnitude of sand moving winds is less, with velocities of only

Table 1. Average slope inclinations of the northern cliff (in degrees) for a 12 m horizontal distance according to Fig. 2-A

Measuring point	Distance (in m) from point #1	Inclinations in azimuth 295°	Inclinations in azimuth 250°	Inclinations in azimuth 235°
1	0	12.95	12.95	7.40
2	20	9.09	14.03	14.57
3	40	26.56	26.68	19.79
4	60	41.34	24.70	17.22
5	80	30.54	24.22	19.29
6	100	34.60	31.30	13.49
7	120	26.56	18.26	9.64
8	140	19.29	11.30	6.27
9	160	11.30	3.43	0.60

Table 2. Average slope inclinations of the southern cliff (in degrees) for a 12 m horizontal distance according to Fig. 2-B

Measuring point	Distance (in m) from point #1	Inclinations in azimuth 295°	Inclinations in azimuth 250°	Inclinations in azimuth 235°
1	0	16.70	14.03	20.80
2	20	31.00	26.56	18.77
3	40	40.60	29.24	20.80
4	60	39.35	27.92	18.77
5	80	38.65	29.68	21.80
6	100	39.35	26.56	20.80
7	120	31.38	20.80	14.03
8	140	29.24	20.80	15.10
9	160	28.36	17.22	14.57
10	180	26.56	11.85	10.75
11	200	16.69	16.17	12.95
12	220	23.74	18.77	7.40
13	240	7.12	26.56	14.03

7 to 13.9 m s⁻¹ during 7.8% of the time. The high magnitude winter winds have a West to South-West direction. Measurements taken of aeolian discharge by sand traps in the 1985/86 winter at Ashdod showed that most of it came from a West and South-West direction [10].

1.5 Research methods

The field work encompassing wind velocity and direction, as well as sand discharge on the cliffs and its surroundings, was carried out during winter storms after an alert on an oncoming storm from the forecasting department of the Meteorological Service. In 7 out of 11 of the anticipated storms, the wind was above threshold velocity and sand movement was monitored.

Wind speed measurements were taken by microanemometers, and wind directions were deduced from the formation of shadow dunes on the lee side of obstacles. Sand discharge was measured by sand traps similar to those developed by Leatherman [12] and Rosen [14].

The filling of sand traps during some of the winter storms took 5 to 19 minutes. This indicates that it is useless to leave traps in the field without supervision under these conditions. In addition, the traps do not rotate and have to be oriented in the field according to the wind direction of the storm.

2 Results

2.1 Windflow on and around the cliffs

All the winter storms on which measurements were taken followed a South-West direction. Winds from this direction are diverted at the foot of the cliff, flows parallel to the cliff front until its northern end is reached, and then deflects eastward (Fig. 4). The impact of the cliff on wind magnitude is depicted in Fig. 5. As the wind approaches the cliff, its velocity increases toward the crest. The slope of the cliff is concave (Fig. 5) with a step at a distance of 20—30 m from the beginning of the cross-section. According to the topography and variations in wind magnitude, it is assumed that at the foot of this step and behind it, the windflow separates, which results in a great abatement in wind magnitude under the upper separation eddy. After the upper reattachment, a gradual increase in wind velocity toward the crest occurs.

Figure 6 shows the magnitude of deflected wind blowing parallel to the foot of the cliff during a storm from 245°. There is a gradual increase in wind velocity up to 15% between points A and B.

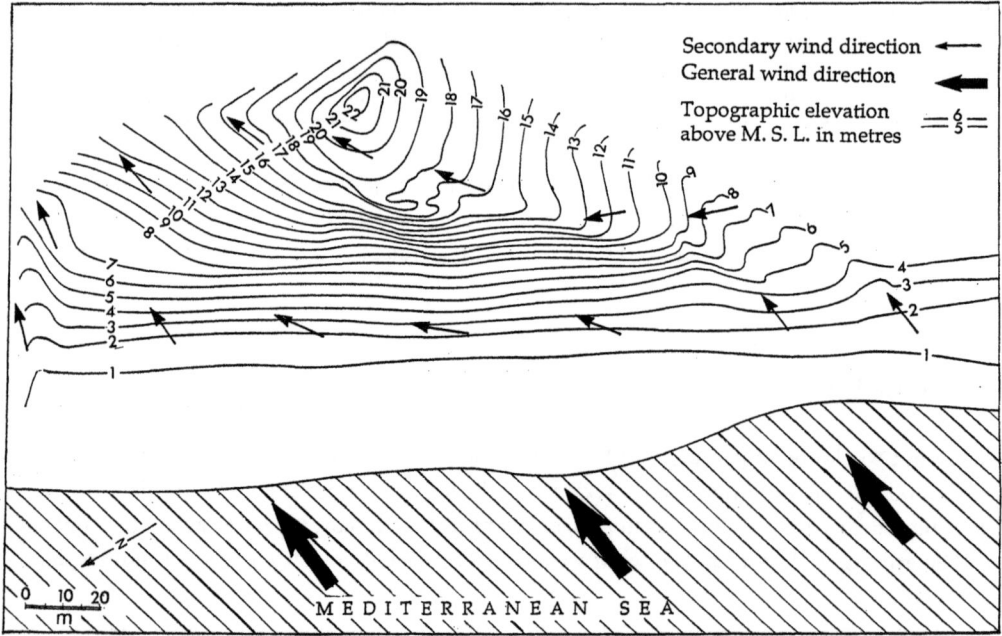

Fig. 4. Map of the secondary windflow on and around the northern cliff during the storm of 12 January 1988

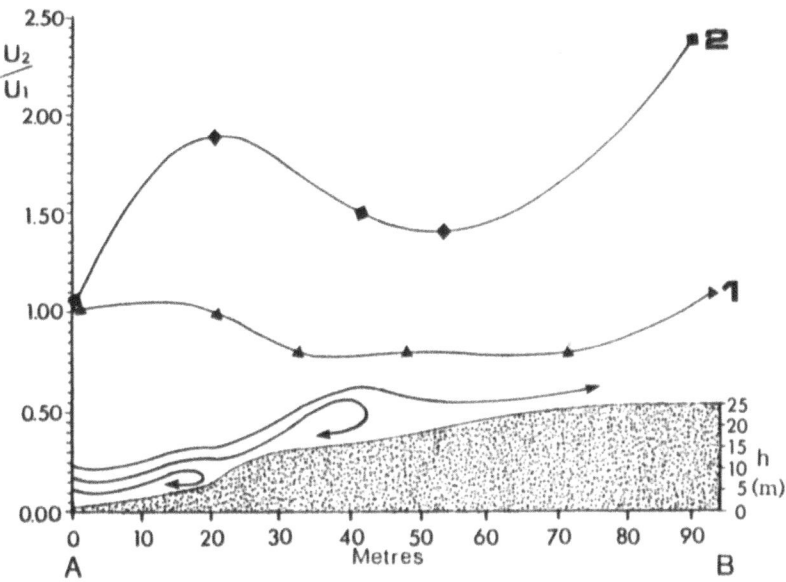

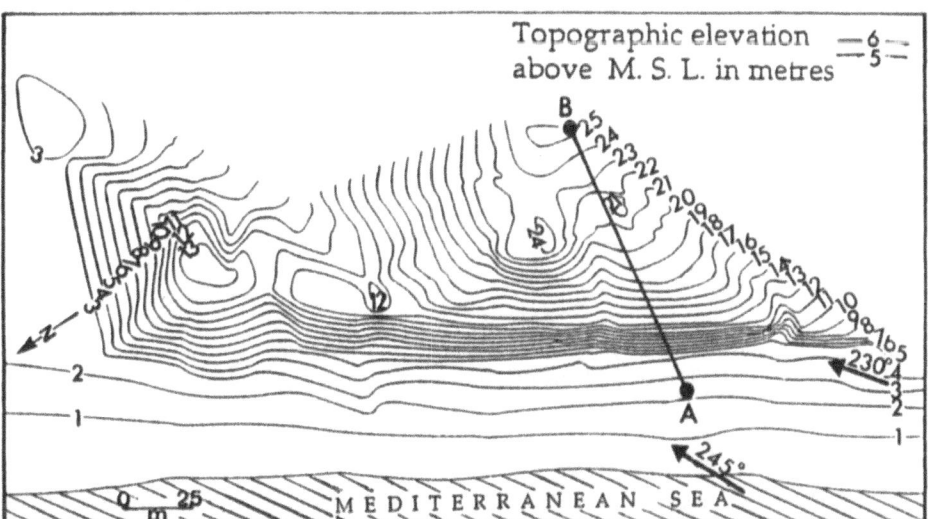

Fig. 5. Changes of wind magnitude along a cross-section (A-B) on the southern cliff under two different wind directions: 1 — 230°, 2 — 245°. U_1 is the wind velocity at 2 m height on the shoreline and U_2 is the wind velocity at 2 m height at various points of the cross-section. The two flow separations, as depicted over the cross-section, are based on the flow along the base of the cliff and the abatement in wind velocity along the cross-section

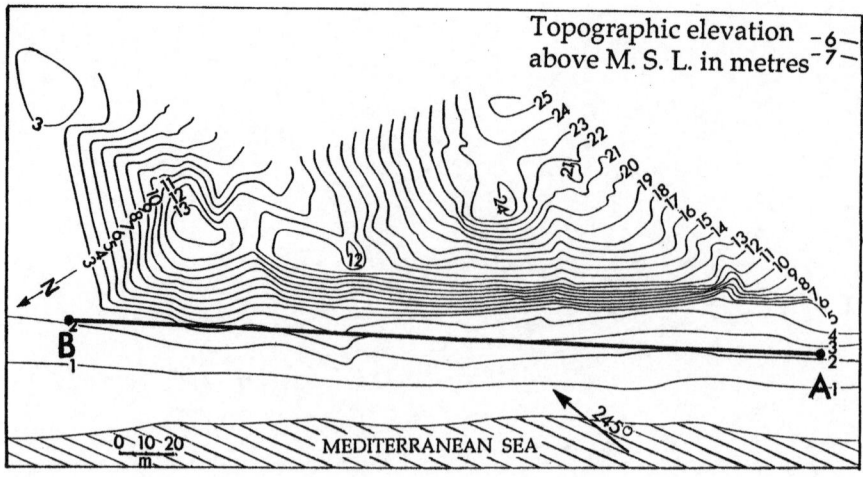

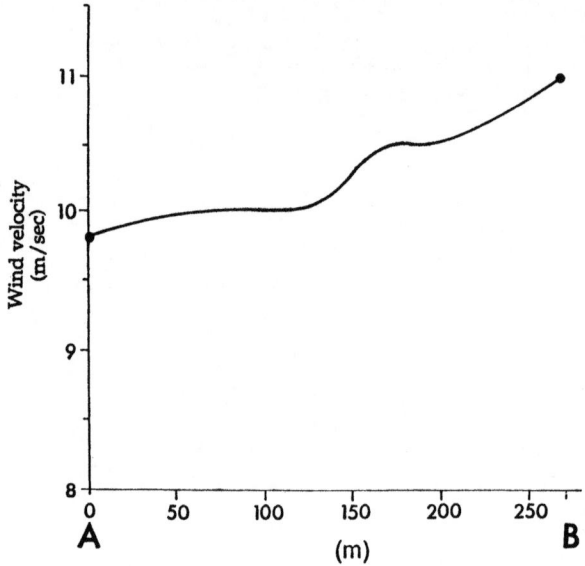

Fig. 6. Changes in the velocity of the secondary wind as measured at 2 m height along the foot of the southern cliff (A-B). The general wind direction during this storm was 245°

2.2 Aeolian sand transport on and around the cliffs

Sand transport measurements were similarly taken during winter storms which were sometimes accompanied by showers that reduced sand transport to practically zero. Rain, on the occasion of the storm of 19 December 1987, elevated sand moisture (in weight) to 1.8%, which annulled sand transport.

Taking into consideration that winter winds blow from the South to West, three traps facing South—West were accordingly deployed on the moderate slope of the cliff (Fig. 7). Results show (Table 3) that sand did not climb even this moderate slope of 6°–7° declination but was transported along the foot of the cliff toward its northern end, where the wind is deflected inland. Hence, a great amount of sand was trapped at the foot of the cliff; but, from there inland, there was a drastic drop in trapped sand, which corroborates that most of the sand is deposited there, forming sand dunes (Fig. 3).

A similar measurement of sand transport was done on the storm of 1 February 1988. This was an exceptional storm, because its wind velocities reached a magnitude of 26 m s⁻¹ (at 2 m high). In addition, several days that preceded the storm were warm and dry and the beach sand was dry and ready for aeolian transportation. Figure 8 shows the deployment of the traps during the storm and Table 4 shows the sand discharge as measured by the traps. Even in this strong dry storm, sand did not climb the moderate south facing cliff. Sand movement only occured along the foot of the cliff and up to a height of 3 m.

These two phenomena can be interpreted with the flow direction (Fig. 4). The southwestern onshore wind which is oblique to the cliff is diverted to flow parallel to the cliff front. Because of that, sand is incapable of climbing the cliff. The sand transported along the foot of the cliff is deposited at the northern end, where the cliff is discontinuous and the flow is deflected inland.

Because sand movement only occured along the foot of the cliffs, two measurements of sand transport were taken there. Figure 9 shows the location of six sand traps positioned along the front of the two cliffs. Figure 10 shows the results of sand discharge during two storms, one of 25 of December 1988 and the other on 9 January 1989. These storms were accompanied by rain which brought the moisture content (by weight) of the sand to values of from 2.3% up to 3.66%. Results show that the sand discharge increases along the cliff, at first, moderately (or even decreasing slightly), and thereafter increasing at a considerable rate and, finally, stabilizing between traps no. 5 and 6.

Table 3. Aeolian sand discharge as measured during the storm of 12 January 1988 on the northern cliff

Trap no. (according to Fig. 7)	Discharge ($m^3\ hr^{-1}\ m^{-1}$)	Wind velocity at 200 cm ($m\ s^{-1}$)
1	0.1550	18
2	0.0966	19
3	0.0027	18
4	0.0002	19
5	0.0000	18
6	0.0000	18
7	0.0000	20

Table 4. Aeolian sand discharge as measured during the storm of 1 February 1988 on the southern cliff

Trap no. (according to Fig. 8)	Discharge ($m^3\ hr^{-1}\ m^{-1}$)	Wind velocity at 200 cm ($m\ s^{-1}$)
1	0.6900	20
2	0.6030	20
3	0.6030	20
4	0.0000	20
5	0.0000	19
6	0.0000	19
7	0.0000	25

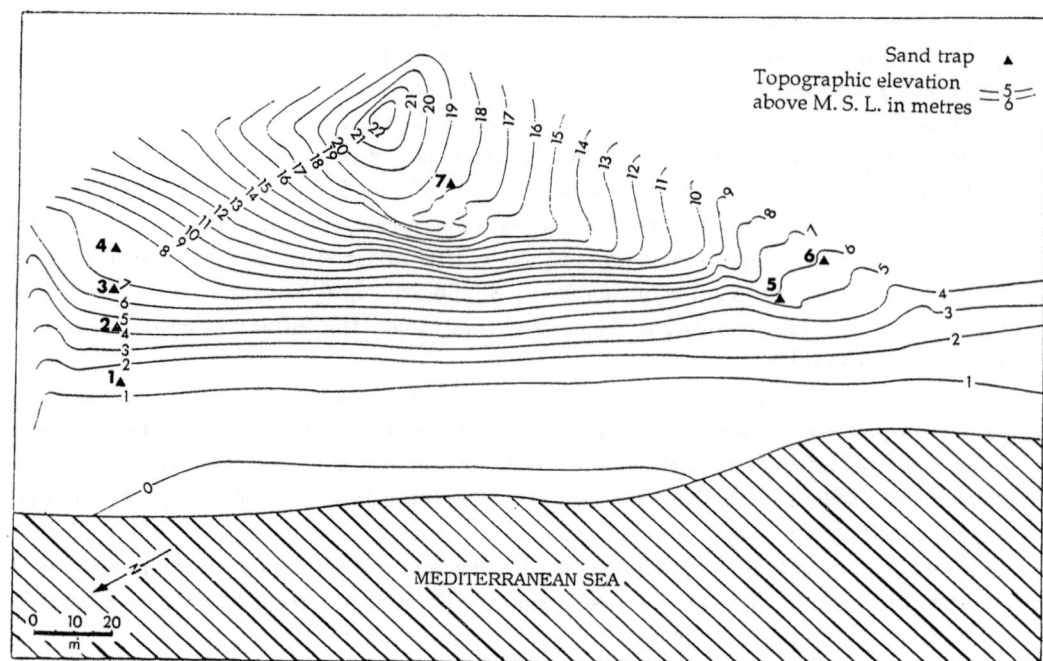

Fig. 7. The locations of the traps on the northern cliff during the storm of 12 January 1988. Compare this Figure with Fig. 4 showing the wind direction in this storm

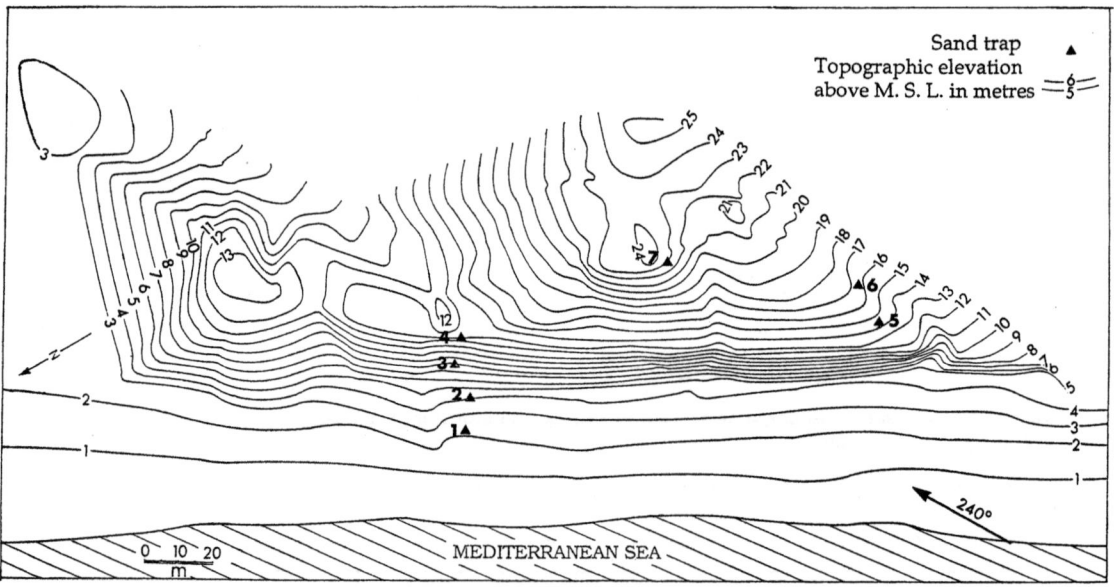

Fig. 8. The locations of the traps on the southern cliff during the storm of 1 February 1988. The general wind direction is shown by the arrow

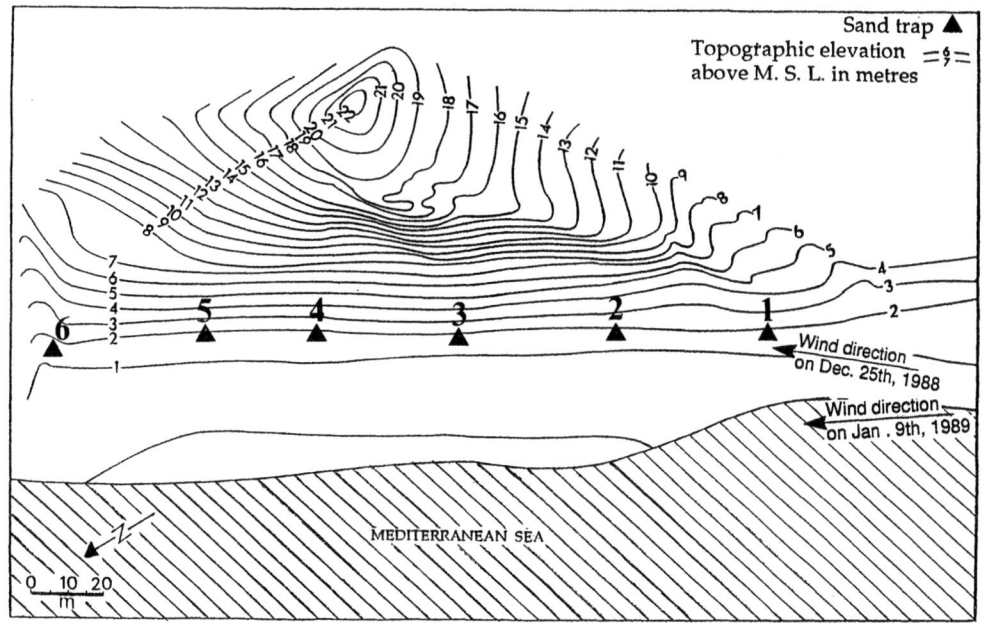

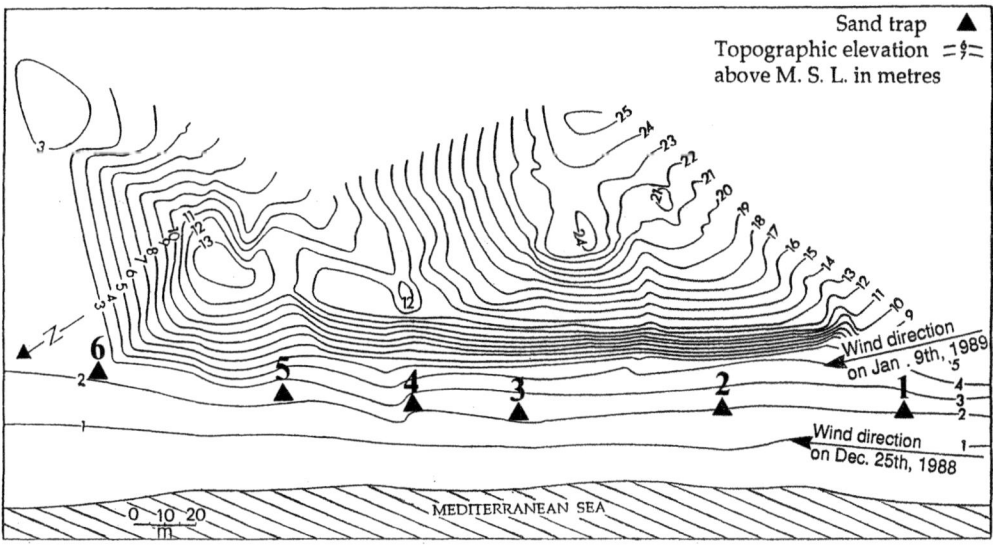

Fig. 9. The location of the traps along the foot of the cliffs during the storms of 25 December 1988 and 9 January 1989. Arrows show the wind direction during the two storms after the divertion of the wind to flow parallel to the front of the cliff. **A** northern cliff, **B** southern cliff

2.3 The change in grain-size in the transition between the beach and aeolian environments

The sand was deposited on the beach by the swash of the waves and was carried from there by the wind. Samples of sand were taken from the beach surface near the traps and from aeolian sand snared by the traps. Grain-size analyses were undertaken by means of standard sieves suspended on a shaker. The orifices of the sieves were between 0.062 and 2.000 mm in 0.25 phi class intervals. Calculations of the results of the four moment statistics of the sieving were performed using the SAHARA software [7].

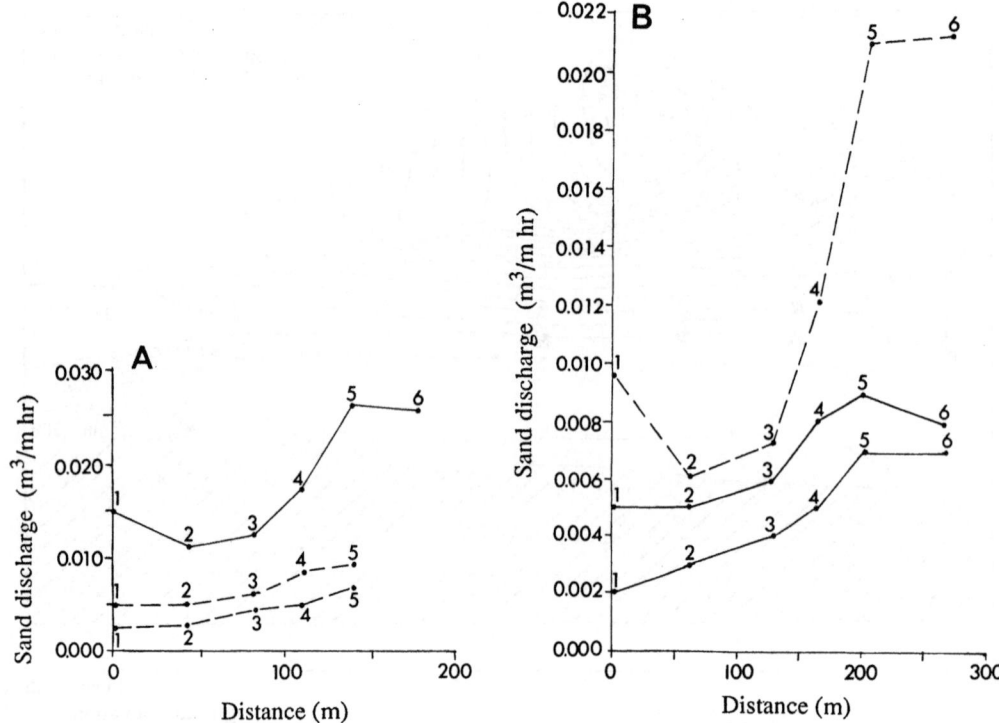

Fig. 10. Changes in sand discharge along the foot of the cliffs during the storms of 25 December 1988 (continuous lines) and 9 January 1989 (dashed lines). **A** northern cliff, **B** southern cliff

Results showed that the grain-size of the trapped sand was in the range of $2-3\varphi$ (0.125 to 0.250 mm), known as the average grain-size of aeolian sand [1][18], and its average (2.24φ) was relatively finer than that of average beach sand found near the traps (1.81φ). The standard deviation of all the samples was in the range of well-sorted sand [8]; however, the standard deviation of the trapped aeolian sand samples was smaller than that of the beach sand. No obvious trend for change was observed in the mean grain-size along the foot of the cliffs, but changes in the mean grain-size of the beach sand did correspond to similar changes in the mean grain-sizes of the trapped sand.

The skewness results did not indicate any particular change along the sand transport path or any significant changes between the beach and the aeolian sand. The kurtosis of the beach sand was negative and platykurtic, while the trapped sand showed positive kurtosis values and a leptokurtic shape.

Statistical t-tests done on the mean and standard deviation values of the samples from the two environments indicated that, except for one case, aeolian sand is significantly different from beach sand. Hence, transport of beach sand by the wind for even a very short distance will unquestionably change the character of beach sand to an aeolian one.

The statistical moments were calculated on the basis of log-normal or near log-normal grain-size distribution although many samples deviated from log-normal grain-size distribution. By using the logarithmic hyperbolic distribution form (shown to model size distribution of aeolian sand [4]), the sand samples proved to be close to a hyperbola. Samples collected on the beach were found to be badly described by a log-hyperbolic distribution. Hartmann [11], who analysed 3000 sand samples from beaches in Israel, found that 30% of them could not be fitted by a log-hyperbolic distribution.

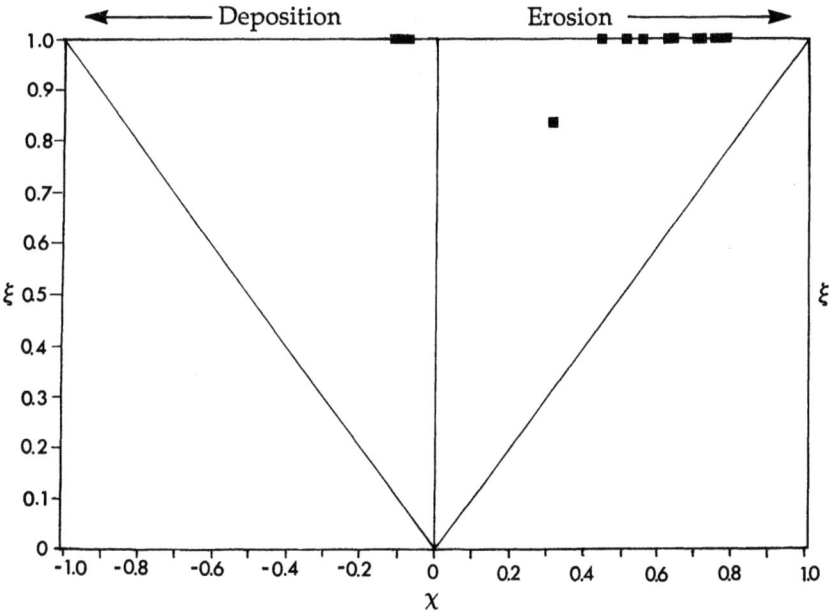

Fig. 11. The position of the skewness (χ) and kurtosis (ξ) values in the hyperbolic shape triangle of the grain-size distribution of the aeolian sand shared by the traps

Samples collected from the traps, however, were well fitted by a log-hyperbolic distribution. Figure 11 shows the domain of variation of the invariant parameter χ and ξ, in a triangle known as hyperbolic shape triangle [5] (χ expresses the skewness and ξ the kurtosis of the hyperbolic distribution). In an erosional environment the position of (χ, ξ) tends to move in the right-hand direction in the hyperbolic shape triangle [5]. Most of the sand samples collected from traps no. 4 to 6 (Figs. 9 and 10) fall within the right side of the hyperbolic shape triangle (Fig. 11). The four samples, which fall in the left side of the triangle, were taken from the first four traps (Fig. 10).

3 Discussion

3.1 The effect of sea cliff on inland encroachment of aeolian sand

The results indicate that under the winter wind directions a cliff with a topographical dimension similar to that of the two cliffs of the research area acts as a total barrier impeding aeolian sand encroachment inland from the beach. Onshore wind approaching the cliff perpendicularly is separated from the surface at a distance that is less than one cliff-height upwind of the cliff [16]. As a flow must be continuous with no holes or voids in it, a countercurrent ensues, which comes down from the cliff onto the separation zone. Thus a separation eddy develops (Fig. 12 A), which prevents sand from climbing the cliff [16].

When the wind impinges on the cliff obliquely, the separation eddy turns into a helical roll vortex [16] (Fig. 12 B) which moves along the cliff front and provokes sand transport along the foot of the cliff (Figs. 6, 9, 10). Formation of transient echo dunes in front of cliffs during storms corroborate this mechanism of a helical roll vortex along the foot of a cliff

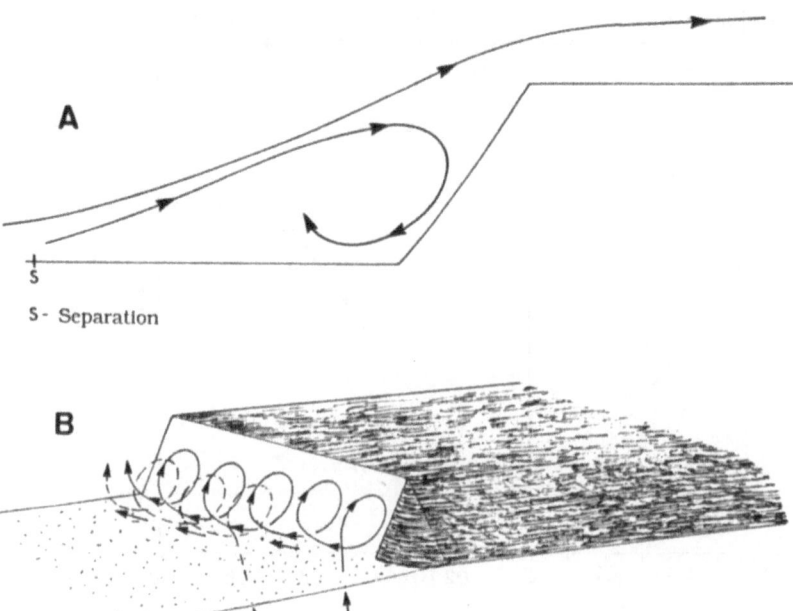

Fig. 12. A model of windflow in front of the sea cliff. **A** the situation in two dimensions, **B** a three-dimensional model in which a helical roll vortex has developed parallel to the cliff front. The arrows show the direction of the sand transport. This figure is based on field observations and wind tunnel simulations [16]

[16]. This model illustrates results from wind direction and magnitude, and sand discharge measurements (Figs. 4, 6, 7, 8, 9, 10; Tables 3, 4).

The helical roll vortex deflects the wind direction along the foot of the cliff, where wind magnitude is accelerated because more streamlines join the helical flow. This leads to an increase in sand transport along the foot of the cliff. On the northern side of the cliff, where a gap is located, another wind deflection takes place and sand penetrates inland. Immediately after this latter deflection, an abrupt abatement in wind velocity brings about deposition of sand (Figs. 7 and 8) and the formation of sand dunes there (Fig. 3). If the cliff had been constant and continuous, as in the Sharon region, sand would not be able to encroach inland from the beach.

3.2 The effect of the fetch on aeolian sand transport on the beach

The increase of sand discharge along the cliff for distance of 150 to 200 m (Fig. 10) can be interpreted as the surface distance (fetch) required for reaching a steady-state which is characterized by a specific total mass flux, an equal number of impacting and ejected grains, and a stationary wind velocity profile [2]. However, a steady state is achieved within 1 to 2 seconds [2]. A fetch value of 9 m was found to correspond to sand movement in wind tunnels [3] [6], while a fetch of 10—20 m was measured on the Dutch beach [15]. It seems probable that the high values of the surface distance required for reaching an equilibrium flow of wind and sand in the present study are the result of the diversion of the wind by the cliff front (Fig. 4) and of the subsequent boosting of its magnitude by the non-homogeneous secondary winds, while flowing along this path (Fig. 6).

4 Conclusions

(1) A cliff with a configuration similar to that which was the object of the present research area, under winter wind directions, may act as a barrier to aeolian sand encroachment from the beach inland. Sand penetrating inland and may only take place through gaps along the cliff.

(2) In Israel, the dominant storm wind direction responsible for aeolian beach sand encroachment is from West to South—West. Under this wind regime, which occurs in winter time, there is always windflow deflection and resulting sand movement along the cliff in the direction of its northern end. At this point the wind is deflected again and carries sand through this gap inland. Alongside this latter deflection, however, the wind magnitude abates and most of the transported sand is consequently deposited. For this reason sand dunes were found at the northern end of each cliff.

(3) Because of windflow deflection, no particular limit for fetch length is set on the beach. However, the effect of the non-homogeneous secondary winds, resulted from the windflow diversion along the cliff, is to increase the sand flux along a distance of 150 to 200 m.

(4) The main aeolian transport of beach sand inland occurs during wintertime during successive storms, where wind velocities (at 2 m hight) may attain a range of above 20 m s^{-1}.

(5) According to aeolian sand discharge measured by sand traps and wind data of the Ashqelon area, estimates are that the minimal aeolian transport of beach sand inland during the four winter months, through the gaps north of the cliffs in the research area is 8.7 m^3 m^{-1}.

(6) Aeolian beach sand encroachment inland is affected by sand moisture. When sand moisture (by weight) reached 1.8%, no sand movement was observed even during a storm of 26 m s^{-1} (at 2 m high). However, in a following storm with wind velocities of 12 m s^{-1} and with a sand moisture above 2%, some aeolian sand transport did take place. From this it is concluded that sand transport after rain may occur when wind blows for long time between one rain event and the next, causing the first few centimetres of the surface to dry up.

(7) Aeolian sand snared in traps is finer and better sorted than beach sand deposited on the beach.

Acknowledgements

We are grateful to D. Hartmann for his help in analysing grain-size data, and to P. Hatokai for his help in the field.

References

[1] Ahlbrandt, T. S.: Textural parameters of eolian deposits. In: A study of global sand seas, (McKee E. D., ed.). Prof. Pap. U.S. Geol. Survey 1052, pp. 21—51, 1979.
[2] Anderson, R. S., Haff, P. K.: Simulation of eolian saltation. Science 241, 820—823 (1988).
[3] Bagnold, R. A.: The physics of blown sand and desert dunes. London: Methuen 1941.
[4] Bagnold, R. A., Barndorff-Nielsen, O.: The pattern of natural grain size distributions. Sedimentology 27, 199—207 (1980).

[5] Barndorff-Nielsen, O. E., Christiansen, C.: Erosion, deposition and size distributions of sand. Proc. Roy Soc. London A417, 335—352 (1988).

[6] Chepil, W. S., Milne, R. A.: Comparative study of soil drifting in the field and in a wind tunnel. Sci. Agr. 19, 249—257 (1939).

[7] Christiansen, C., Hartmann, D.: SAHARA: a package of PC-computer programs for estimating both log-hyperbolic grain size parameters and standard moments. Computers Geosci. 14, 557—625 (1988).

[8] Folk, R. L., Ward, W. C.: Brazos River bar: a study in the significance of grain size parameters. J. Sed. Petrol. 27, 3—26 (1957).

[9] Goldsmith, V., Golik, A.: Sediment transport model of the southeastern Mediterranean coast. Marine Geol. 37, 147—175 (1980).

[10] Goldsmith, V., Rosen, P., Gertner, Y.: Eolian sediments transport on the Israeli coast. Final Report, U.S.-Israel BSF, National Oceanographic Institute, Haifa, 1988.

[11] Hartmann, D.: Coastal sands of the southern and coastal part of the Mediterranean Coast of Israel — Reflection of dynamic sorting processes. Unpubl. Ph. D. dissertation, Aarhus University, 1988.

[12] Leatherman, S. P.: A new aeolian sand trap design. Sedimentology 25, 303—306 (1978).

[13] Nir, Y.: Sedimentological aspects of the Israel and Sinai Mediterranean coasts. Geol. Survey of Israel, Report 39/88 (1989).

[14] Rosen, P. S.: An efficient, low cost, aeolian sampling system. Scient. Tech. Notes, Current Res. Part A, Geol. Surv. Canada, Pap. 78—1A, 531—532 (1979).

[15] Svasek, J. N., Terwindt, J. H. J.: Measurements of sand transport by wind on a natural beach. Sedimentology 21, 311—322 (1974).

[16] Tsoar, H.: Wind tunnel modeling of echo and climbing dunes. In: Eolian sediments and processes, (Brookfield, M. E., Ahlbrandt, T. S., eds.), pp. 247—259. Amsterdam: Elsevier 1983.

[17] Tsoar, H.: Trends in the development of sand dunes along the southeastern Mediterranean coast. Catena [Suppl.] 18, 51—60 (1990).

[18] Tsoar, H.: Grain size characteristics of wind ripples on a desert seif dune. Geogr. Res. Forum 10, 37—50 (1990).

Authors' address: Dr. H. Tsoar and Mr. D. Blumberg, Department of Geography, Ben-Gurion University of the Negev, Beer-Sheva, IL-84105, Israel.

Acta Mechanica (1991) [Suppl] 2: 147—159

Coastal erosion and aeolian sand transport on the Aquitaine coast, France

J.-M. Froidefond and R. Prud'homme, Talence, France

Summary. The coastal dunes of Aquitaine number more than 1 500, and have a total included volume of between 10 and 20×10^9 m^3. There are four major types of dunes: parabolic dunes, barchans, round dunes and the foredune. All the dunes were constructed by westerly winds, blowing off the sea. Dating of these dunes is based on paleosols of the Great Dune of Pilat, the highest dune in Europe. The base of this dune is dated at about 3 500 years B.P., but most of the dunes were probably formed during the last 2 000 years. These dunes obstructed a number of small rivers inducing the development of lakes lie behind them. One small estuary was transformed into a lagoon, the Arcachon lagoon.

The sands which built the dunes come from the Atlantic coastal beaches. The present rate of coastal retreat is about 1 to 2 m a year. The coastal erosion rates now observed can be used to estimate the rate of intrusion of sand inland from the coast. Measurements with sand traps and by cartographic comparison give an eastward aeolian flux estimated of between 15 and 30 m^3 a year per m length of foredune.

Introduction

The western part of the Aquitaine region is covered by sands in continental and coastal dunes. This region, planted by pines (the largest pine forest in Europe) during the XIX century [20], may be subdivided into two areas, as follows:

1. A great plain called "Les Landes de Gascogne", the sandy moor of Gascony (Fig. 1) contains numerous continental dunes, generally parabolic dunes and sand ridges [11], [16]. Their asymmetrical shape implies winds blowing from the Atlantic Ocean. For the parabolic dunes, the best known example is the Cazalis dune located at the edge of the "Grandes Landes" about 10 km NNW of Captieux. These continental dunes are interpreted as pre-Holocene from exoscopic examination of quartz [16].

2. The coastal dunes are located between the Atlantic sandy shore and the lakes of Gascony (from north to south, lakes Hourtin, Lacanau, Cazaux, Biscarrosse, Mimizan, Leon and Souston). In this area (Fig. 1) which is 8 km wide and 230 km long, there are many dunes of different types. Since the end of the XIX th century [20], these coastal dunes have been covered by a pine forest.

These dunes have been studied to explain the relations between aeolian sand transport and coastal erosion processes, a problem debated by several researchers: [2], [5], [12], [14], [18]. To present this subject for discussion it is necessary to examine:

— the characteristics of coastal dunes (shapes, frequencies, volumes, winds and results of sand trap experiments);

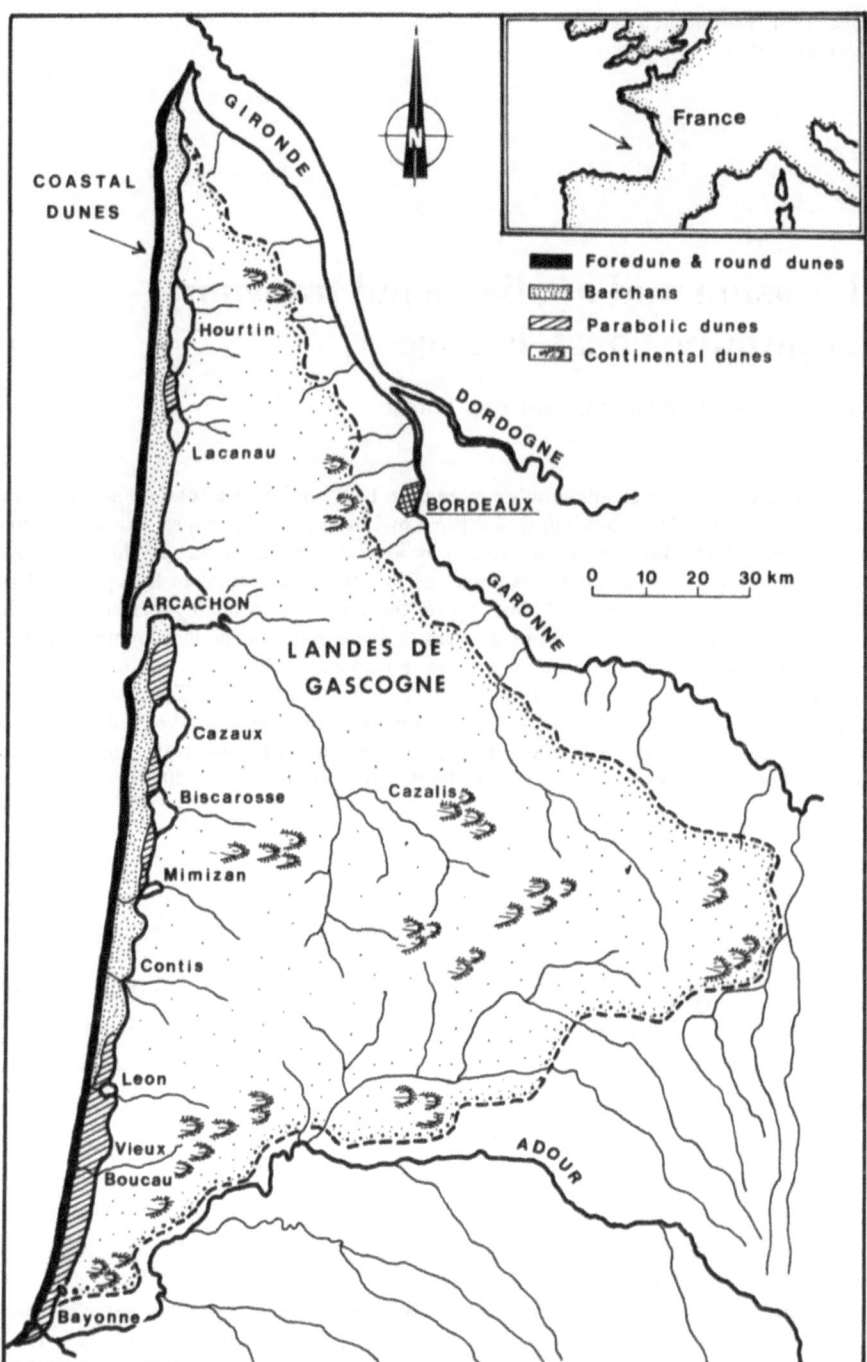

Fig. 1. The continental and the coastal dunes of the Aquitaine. The coastal dunes stretch from the shore to the line of lakes. The coastal dunes (except the foredune) and the Landes de Gascogne are covered by a pine forest

— the chronology of deposition from the Great Dune of Pilat example,

— the coastal erosion and the effect of sea level variations on coastal dune development
and the present aeolian flux.

1 Coastal dunes of Aquitaine

1.1 Morphology of the coastal dunes

The coastal dunes of Aquitaine can be classified into four categories in terms of morphology.
These categories are (from east, inland, to west, Atlantic Ocean): — parabolic dunes —
crescentic or barchan dunes — round dunes — foredune or aeolian sand barrier along
the shore (Fig. 1). All these dunes show asymmetrical morphology, with abrupt slopes leeward to the east and gentle slopes windward to the west. Clearly they have been shaped by
westerly winds. These coastal dunes cover the soil of the sandy moor of Gascony whose
altitude rises up from 0 m (along the coastline) to 12 m inland, at 8 km away (lakes line).

— The parabolic dunes correspond to the first coastal individualised dunes. They were
covered by pine forest before the XVIIth century. These dunes are located between the
lakes (Fig. 1). Towards the south, they form the great majority of coastal dunes. Generally
their altitudes are less than that of the barchans.

— The barchans cover an area continuous from the Gironde Estuary to Moliets. Sometimes they coalesce to form great ranges of dunes (Fig. 1).

— The round dunes, also called chaotic dunes, are smaller than the barchans and the
parabolic dunes, usually less than 20 m in height. They are located between barchan and
foredune areas. These dunes have probably developed recently in the presence of vegetation.

— The foredune, an aeolian sand barrier behind the sandy beach, has been greatly
promoted since 1890 by works made to stop sand drift (i.e. palisades and the planting of
grass (Ammophila arenaria). Unlike the barchan and parabolic dunes, which are fixed by
forest, the foredune has a character which changes in terms of wind velocities and grass
cover. This foredune is actively managed by the "Office National des Forêts" (National
Forest Agency) by means of permanent workers who reprofile and replant grasses to reduce
the aeolian drift and to protect the pine forest [4].

1.2 Number of dunes, sand volume and wind orientation

In terms of amplitude, the whole area of coastal dunes includes:

    1 dune above 100 m height, The Great Dune of Pilat (or Pyla),
   47 dunes between 70 and 100 m in height
  350 dunes between 50 and 70 m in height
  710 dunes between 30 and 50 m in height
  410 dunes between 10 and 30 m in height.

So in this area, between the sandy shore and the lakes, 8 km wide and 230 km long,
they are over 1500 dunes with amplitudes of more than 10 m. The total volume of these
coastal dunes may be estimated at between 10 and 20×10^9 m^3 [14], [16].

Observation of the orientation of dune axes may give information on the direction of

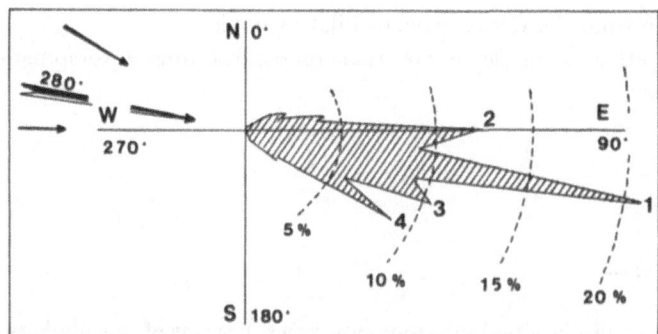

Fig. 2. Direction of dominant winds from the study of coastal dune axes. The dominant directions are westerly and north-westerly. Maximum N 280° (1) N 270° (2) N 290° (3) N 300° (4) (from Legigan [16])

the wind shaping the dunes. Results of about 600 observations [16] reveal that coastal dunes have been shaped by westerly winds (essentialy N 280°) without appreciable changes between barchans and parabolic dunes (Fig. 2). These results suggest a steady wind climate between the parabolic and barchan dune building periods.

1.3 Sedimentology of the dune sands

The coastal dunes consist predominantly of siliceous grains with 1 to 2% of heavy mineral grains. Granulometric analysis showed that the parabolic dune sands are well sorted and have a mean size of about 0.3 mm and contain a small (< 5%) coarse fraction. The barchan dune sands are very well sorted with a mean grain size of about 0.35 mm [2].

The mean grain size of the foredune changes from north to south. The mean size decreases from 0.35 mm near Soulac to reach a minimum of 0.25 mm near Arcachon. Farther south, the mean grain increases progressively and reaches 0.40 mm near Bayonne.

The beach sands are generally coarser than the dune sands with a mean grain size of between 0.35 and 0.5 mm and with gravels at the seaward end of the intertidal zone. The inner continental shelf, between 0 and 50 m depth, is covered by sands, coarse sands and gravels [1].

2 Dating of coastal dunes and of the Great Dune of Pilat

The Great Dune of Pilat has been studied to date the invasion of barchan and parabolic dunes. This great dune is located along the Arcachon lagoon channel (Fig. 3).

2.1 Dunes of the Arcachon lagoon area

Topographic maps and aerial photographs from Institut Geographique National (I.G.N.) were used to map the dune fields. From north to south, they are the following (Fig. 3):

— on the Cap-Ferret spit, barchans or crescent dunes of 20 to 40 m in height and a foredune along the coast with very active barchans (shore dunes);
— to the South of Arcachon city, there are some great transverse dunes and crescent dunes. Between these great dunes and the Cazaux lake, one finds a parabolic dune field. The parabolic dunes are partly over-run by the barchan field (Fig. 3).

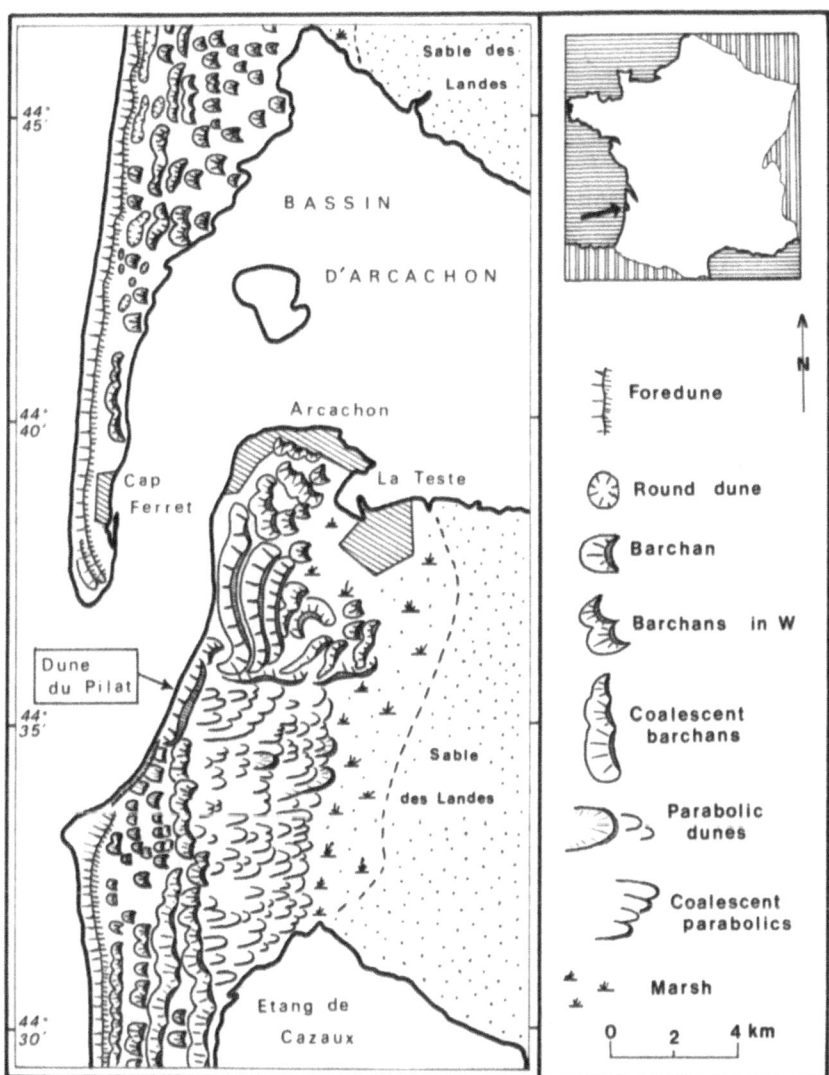

Fig. 3. The coastal dunes on both sides of the Arcachon lagoon from the study of aerial photographs. All the dunes are covered by a pine forest except the foredune and the Dune du Pilat

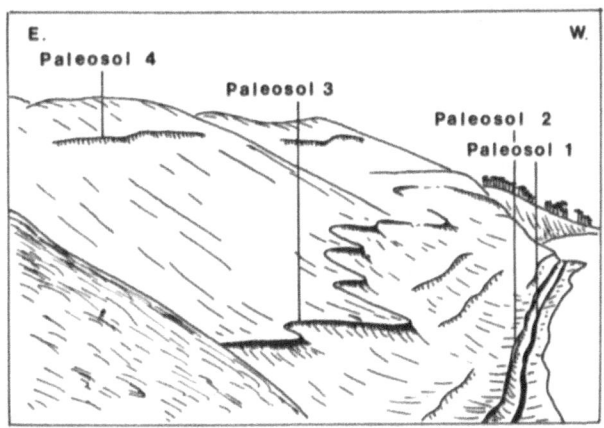

Fig. 4. The Great Dune of Pilat view from the north. Four main paleosols appear on the western side open to the sea winds

Table 1. Paleosol ages

Paleosol	C14 yr. (B.P.)	Palyn. dat.	Hist. dat.	Interpretation	
				Start	End
1	3680 ± 110	Boreal/Atlantic		10000	→ 3500 B.P.
	3460 ± 70				
2	2980 ± 110	Atlantic	Bronze age	3500	→ 2700 B.P.
	2690 ± 70				
3		Recent	XVIth cent.	?	400 B.P.
4			XIXth cent	150 (?) →	100 B.P.

2.2 Morphology of the Great Dune of Pilat

This great transverse dune is 2500 m long and 500 m wide and reaches a height of 105 m (Fig. 4). Its volume is approximately 60×10^6 m^3. The slope of the east side is between 30 and 40° and the slope of the West side between 5° and 20°. This dune progressively moves inland, the east side invading the pine forest. The present rate of translation is between 1 and 5 m per year. Storms accelerate this movement.

2.3 Paleosols and dates

The western side (Fig. 4) shows 4 main paleosols (an old podzol and 3 dune paleosols) plus many heavy mineral beds (several millions) from base to summit of the dune. The dates are indicated by radiocarbon ages [15], palynologic analysis [17], [19] and historic datings [10], (Table 1).

— *Paleosol 1:* The base of the dune is underlined by an old podzol. We can see on this old ground, some stumps of pine. The radiocarbon ages indicate the date of its burial by sand dunes, since 3500 years ago.

— *Paleosol 2:* This one is located between 2 and 5 m above sea-level. The radiocarbon ages are confirmed by some tools dating from the middle Bronze age (VIIth century B. C.).

— *Intermediate paleosols and layers of diatoms:* Above paleosol 2, about 20 m above sea-level, appear 2 to 4 thin dune paleosols. Some are composed of vegetation remains, and others by a layer of siliceous fresh water diatoms (white levels). These diatoms were transported by wind, probably from a pond located in the vicinity [15].

— *Paleosol 3:* The altitude of this paleosol varies from 20 to 40 m. This is the upper part of parabolic dunes. On this surface, we have discovered old coins dating from the XVIth century.

— *Paleosol 4:* Paleosol 4 represents the surface of the "Dune de la Grave" mapped in 1863. The "Dune de la Grave" was a transverse dune reaching 80 m in height, covered by a pine forest planted during the XIXth century to stabilize the aeolian sands, to exploit the wood, and to produce pine resin for industrial use.

— *Summit of the dune:* Paleosol 4 was subsequently buried by 20 to 30 m of aeolian sands derived from the west which formed the "Grande Dune du Pilat" by the beginning of the XXth century. The stacked dune sequence, separated by paleosols, is shown diagrammatically in Fig. 5.

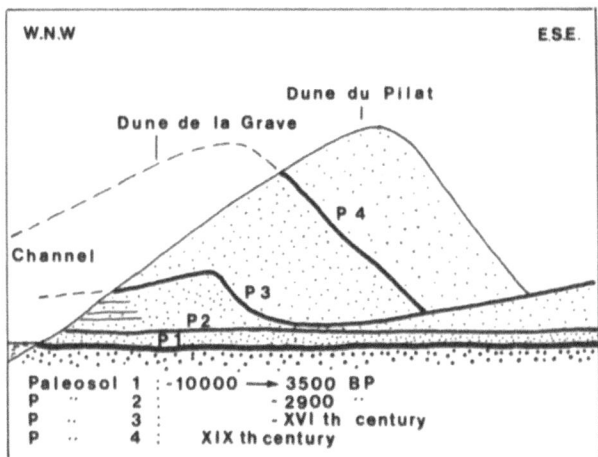

Fig. 5. Cross section showing the geometric disposition of the paleosols

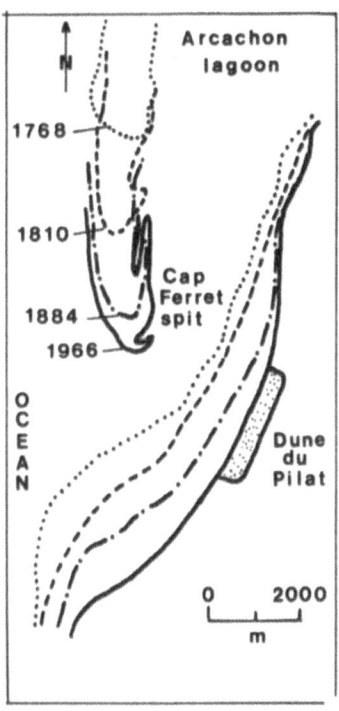

Fig. 6. The inlet of the Arcachon lagoon migrates southward du to the lengthening of the Cap-Ferret spit which induces severe erosion of the western side of the Great Dune of Pilat

2.4 Construction of the Great Dune of Pilat

This is a recent dune built on parabolic dunes after the XVIIth century and essentialy at the end of the XIXth century by accumulation of aeolian sands on a transverse dune, the "Dune de la Grave". This local construction of sand has been generated by the displacement of the Arcachon lagoon channel towards the southeast [9]. The channel migration instigated severe erosion on the eastern shore (Fig. 6). The west side of the "Dune de la Grave" was eroded by westerly winds [9] and sands were carried over the crest and deposited on the lee side.

3 Aeolian accumulation as a consequence of coastal erosion

The preceding results of geomorphologic and geochronologic analysis make possible the following statements:

— The volume of sand dunes in the coastal strip is estimated between 10 and 20×10^9 m³.
— The dynamic factor of aeolian formation is the westerly wind.
— The sand dunes were constructed in the period since 3500 yr B.P. and mainly since 2000 yr B.P.
— Present aeolian processes have a very important role in the formation of the foredune.

To understand how these sand dunes were constructed and what is the source of the sands, complementary results must be presented. In particular, two matters will be developed: the coastal erosion induced by the displacement of the palaeocoast and the contemporary processes.

3.1 Palaeocoast and coastal erosion since 5000 years B.P.

— *Localisation of the palaeocoast*. Previous studies have shown that the palaeocoast at 5000 years B.P. was located to the west of the present shoreline [6], [12], [17]. The curve of sea-level rise (Fig. 7) established by Ters [21] gives an approximation of the sea-level for the last 5000 years. This curve varies from − 7 m (5500 yr B.P.) to 0 m relative to the present lowest level tide (French hydrographic reference, the present tidal range is about 4.5 m in the lagoon). Palaeoenvironment analysis based on micropaleontologic studies, and datings [7] obtained by radiocarbon analysis confirm this curve of sea-level rise.

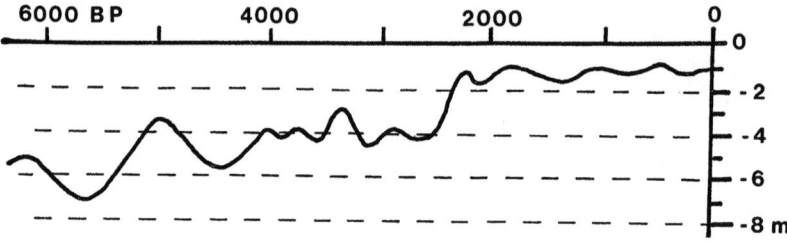

Fig. 7. Sea-level rise since 6000 yr. B.P. (from Ters [21])

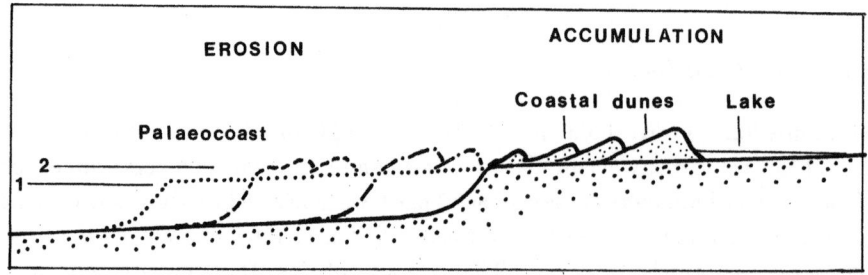

Fig. 8. The relationship between coastal erosion and aeolian accumulation under the influence of westerly winds

To locate the former shoreline at about 5000 yr B.P. we have assumed that the-sea-level was about — 5 m (relative to the present lowest level tide) and that the Aquitaine plaine has the same slope near the coast and inland. The intersection between this slope and the — 5 m horizontal sea-level gives the location of the palaeocoast (Fig. 8) which, on these assumptions, would be situated between 5 and 3 km offshore to the north of Arcachon lagoon and between 3 and 1.5 km to the south.

— *Consecutive erosion and accumulation.* Through the use of theoretical models, it is possible to calculate the volume of sediment displaced by this erosion and that which accumulated as coastal dunes. The coastal area chosen for this experiment is a zone 13 km long between the Adour River and the Capbreton canyon. (The Capbreton study showed the absence of sandy sediments in the head of the canyon [13].) The theoretical volumes of recent sediment accumulations on the inner shelf were calculated by comparing the present surface with one that represents the general shape of the inner shelf (Fig. 8). Then the volume of sediment displaced by transgression was estimated by calculating the volume which accumulated at the foot of the paleocoast located 3 km offshore and dated at 7500—5000 B.P. by Carbonel et al. [6]. The results obtained are as follows:

Volumes of sediments moved by marine erosion: 250×10^6 m³

Volumes of sediments accumulated offshore: 16×10^6 m³

Volume of sand dunes: 300×10^6 m³.

The arithmetic is unbalanced; one notes a deficit of 65×10^6 m³. Therefore it must be concluded that the paleocoast was backed by a substantial dune environment [12]. The results show the importance of aeolian transport and the lesser importance of littoral transport (longshore current and return flow).

3.2 Present erosion and sandy aeolian flux

— *Historic and present erosion.* Comparison of cadastral plans [18], historic maps [14], topographic profiles and the distances between Second World War blockhouses and the shoreline [2] were used to evaluate recent coastal erosion. The following results were obtained:

Areas of erosion are more irregularly scattered. The high erosion areas are separated by about 50 km from one another. The mean retreat exceeded 2 m/y. between Hourtin and Lacanau, Le Porge and Le Grand Crohot, Cap Ferret and to the south of Arcachon, between Mimizan and Contis, and near Capbreton (Fig. 9). In the remaining areas the erosion is about 1 m/y., excepting near Contis and Saint Girons and to the north of Capbreton where the coast is prograding at a rate of up to 4 m/yr.

Thus the recent displacement of the Aquitaine shoreline is irregular with erosion rates varying from 1 to 2 m/y. and a few sectors in accumulation.

— *Measurements of aeolian sand flux.* Sand traps were used over a period of 5 months to measure the volumes of sand transported by the westerly winds on the top of the foredune near La Salie. In order to collect the sand transported by rolling and saltation the back of this trap is located 1.25 m downwind from the entry point (Fig. 10). The collector, which has a volume of 112 dm³, is buried in the ground. The width of the inlet is 10 cm [8]. We have observed that this trap is filled quickly during a wind storm. In this case, the volume collected is under-estimated.

    In spite of these limitations, it is possible to give two extreme values of the volume of
sand transported by the westerly wind into the trap. The minimum value recorded was
800 dm³ for a ten centimeters entry width over a 5 month period. The highest value,
1200 dm³ is the result of calculations based on theoretical volumes calculated from Bagnold
equation [3] during "stormy periods" plus measures in the field during "calm periods".
If we extrapolate these results for one year and on a one metre section, we have a range of
aeolian flux values of 19—29 m³/year/m. However these values cannot be considered to
represent a volumetric balance since the volumes of sand drifted by the easterly wind are
not taken into account [8].

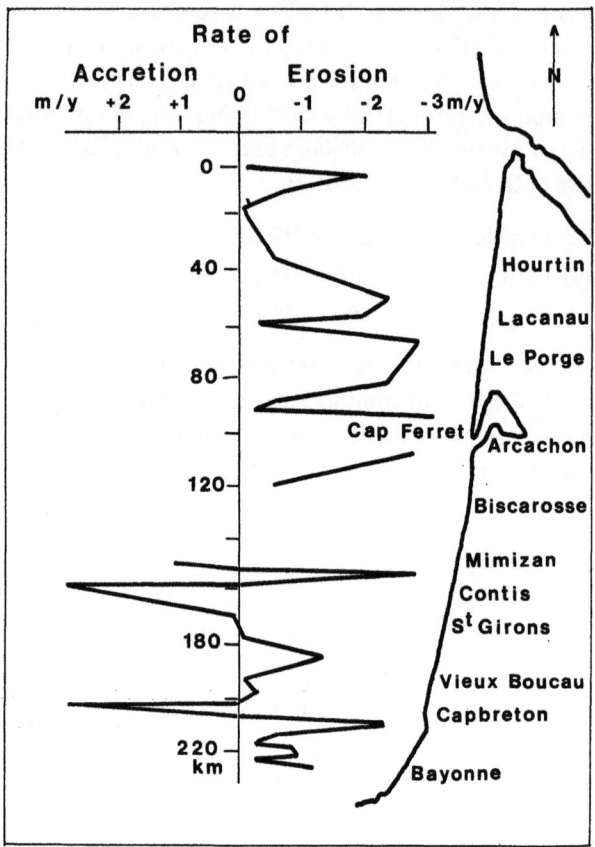

Fig. 9. Recent rate of retreat of the Aquitaine coast between December 1967 and February 1979
(from Migniot and Lorin [18])

— *Flux calculated from the volume of the foredune.* During the end of the XIXth century,
the building of fences by the administration of the Forest Agency (Office National des
Forêts) caused the formation of a foredune parallel to the shore. To calculate the aeolian
flux, the topographic maps of 1875 (coast without foredune near La Salie) and 1966 were
compared. The volume of the foredune in this area was divided by its length, then by the
number of years. Taking into account the innacuracies of topographic contours, the aeolian
sand flux is estimated at between 20 and 40 m³/year/m to the time at which the equilibrium
size was reached. This result accords roughly with flux measurements obtained with the
sand trap.

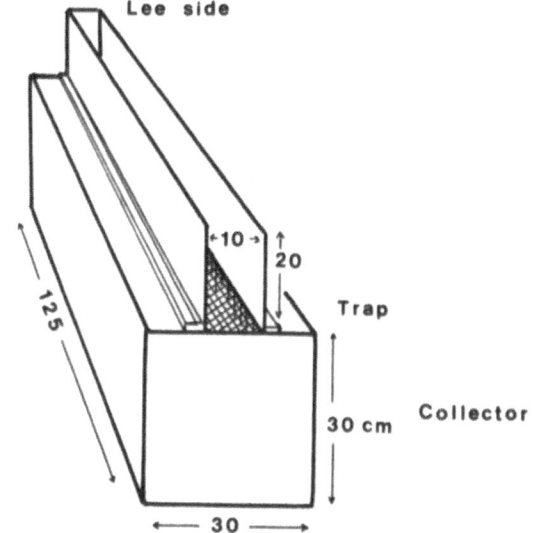

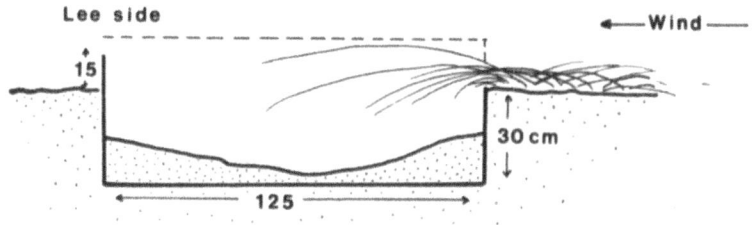

Fig. 10. Sand trap design used to measure the volume of sands displaced by westerly winds on the foredune

3.3 Coastal sand flux in the last 3500 years

Taking the date of origination of the dunes to be 3500 years ago (A), the length (L) of the coast (230 km) and different estimates of the entire volume (V) of sand in the coastal dunes (10 to 20×10^9 m³) it is possible to calculate approximately a mean long term coastal sand flux:

Aeolian flux (m³/y./m) $= V/L/A$

$V = 10, 15, 20 \times 10^9$ m³ $L = 230000$ m $A = 3500$ y.
$V = 10 \times 10^9$ m³ $\rightarrow$ 12.4 m³/year/m
$V = 15 \times 10^9$ m³ $\rightarrow$ 18.6 m³/year/m
$V = 20 \times 10^9$ m³ $\rightarrow$ 24.8 m³/year/m.

These values are similar to those obtained by the other methods described above. Therefore there is a good agreement between coastal erosion and aeolian transport on the Aquitaine coast.

Conclusion

1 — The strong westerly winds blow sands inland and contribute to the rate of coastal retreat which is 1—2 m/year to the north of the Arcachon lagoon and 0—1 m/year to the south.

2 — The present measured rate of sand flux (15—30 m³/m/y.) is sufficient to explain the magnitude of the aeolian deposit (10—20 × 10⁹ m³) formed since 3500 years B.P.

3 — Four generation of coastal dunes exist, from the oldest to the youngest, they are: parabolic dunes, barchans, round dunes and the foredune. Only the parabolic dunes were covered by pines before planting by the Forest Agency. Parabolic dunes (deflation shapes) are probably old barchans changed to parabolic shapes during a phase of semi-stabilization (humid period). The migration of the second phase of barchans blocked a number of coastal streams and generated several lakes behind the dunes. The barchans were planted with pines during the XIXth century. The foredune, an aeolian transverse barrier along the beach, is the result of sand fixation by fences and grasses since the beginning of the XXth century.

4 — To the south of the Arcachon lagoon, near Saint Girons, the coast is advancing seaward at a rate of 1 to 4 m/yr. This progradation probably results from submarine sand flux coming from the North by longshore currents and accumulation in this area. Another area of submarine deposition appears off the Arcachon lagoon inlet.

These sedimentary displacements are important for coastal planning. In order to avoid incorrect decisions, there is an urgent requirment to measure, with more precision, the aeolian and submarine sediment flux in this coastal zone.

Acknowledgements

This study has been supported by the Department of Geology and Oceanography of the University of Bordeaux 1. The authors would like to thank the Organizing Commitee of the Workshop and particularly Dr. B. B. Willetts and the Department of Engineering, University of Aberdeen.

References

[1] Allen, G. P., Castaing, P.: Carte de répartition des sédiments superficiels sur le plateau continental du Golfe de Gascogne. Bull. Inst. Géol. Bassin d'Aquitaine, Bordeaux, 21, 255—261 (1977).

[2] Amini, M.: Etude des processus dynamiques et de l'évolution sédimentaire sur la côte sableuse d'Aquitaine. 373 p. Thèse Doct. Etat., Univ. Bordeaux I (1979).

[3] Bagnold, R. A.: The physics of blown sand and desert dunes. London: Methuen, p. 265 (1941).

[4] Barrère, P.: Memento technique des Dunes du littoral Aquitain. 25 p., Office National des Forêts, Bordeaux, (1986).

[5] Buffault, P.: Sur la Configuration des dunes de Gascogne. 464 p. Delmas Edit., Bordeaux (1942).

[6] Carbonel, P., Duplantier, F., Turon, J. L.: Mise en évidence d'un paléorivage vers — 30 m sur le plateau continental de la région de Capbreton (Golfe de Gascogne). Bull. Inst. Géol. Bassin d'Aquitaine, Bordeaux, 13, 127—143 (1977).

[7] Carbonel, P., Carruesco, C., Pujos, M., Saubade, A. M., Cuignon, R., Fenies, H., Faugeres, J. C.: Mise en place et évolution des milieux de la partie interne du Bassin d'Arcachon. Bul. Inst. Bassin d'Aquitaine, Bordeaux, 42, 5—22 (1987).

[8] Cherbonnier, J., Dedieu, P., Froidefond, J. M., Neviere, E.: Construction d'un piège à sable et mesure du transit éolien sur le littoral Aquitain à La Salie. 146−159, Aderma Edit., In: Actes Coll. Bordomer **85**, Bordeaux (1985).

[9] Clavel, M.: Notice sur le Bassin d'Arcachon. 79 p. Imprimerie Nationale, Paris, In: "Ports Maritimes de la France" (1887).

[10] Dautant, A., Jacques, Ph., Lesca-Seigne, A., Seigne, J.: Découvertes protohistoriques récentes prés d'Arcachon (Gironde). Bull. Soc. Préhistorique Fr. **80** (6) 188−192 (1983).

[11] Enjalbert, H.: Les Pays Aquitains. Le modelé et les sols. 600 p. Imprim. Bière. Bordeaux (1960).

[12] Froidefond, J. M., Duplantier, F., Weber, O.: Estimation and flux of the theoretical sedimentary volume displaced during the Holocene transgression on the Bayonne inner shelf (S. W. France). Marine Geol. **46**, 101−116 (1982).

[13] Froidefond, J. M., Castaing, P., Weber, O.: Evolution morphosédimentaire de la tête du canyon de Capbreton d'aprés les cartes de 1860 et de 1963, utilisation des méthodes informatiques. Bull. Soc. Géol. France, **7** (25) n°5, 705−714 (1983).

[14] Froidefond, J. M.: Méthode de géomorphologie côtière. Application à l'étude de l'évolution du littoral Aquitain. n°18, 273 p. Mémoire Inst. Géol. Bassin d'Aquitaine, Bordeaux (1985).

[15] Froidefond, J. M., Legigan, Ph.: La Grande Dune du Pilat et la progression des dunes sur le littoral Aquitain. Bull. Inst. Géol. Bassin d'Aquitaine, Bordeaux, **38**, 69−79 (1985).

[16] Legigan, Ph.: L'élaboration et la formation des sables des Landes, dépôt résiduel de l'environnement sédimentaire Pliocène-Pleistocène centre Aquitain. 428 p. Thèse d'Etat, Univ. Bordeaux 1 (1979).

[17] Marambat, L., Roussot-Larroque, J.: Paysage végétal et occupations humaines sur la côte Atlantique: l'exemple de la lède du Gurp. Bull. Ass. Française pour l'Etude du Quaternaire, **2**, 73−89 (1989).

[18] Migniot, C., Lorin, J.: Evolution du littoral de la Côte des Landes et du Pays Basque au cours des dernières années. 629−661. Actes du Congrés de Bayonne 28−29 oct. Soc. Sci. Lettres et Art de Bayonne, n. spécial (1978).

[19] Paquerau, M., Prenant, A.: Note préliminaire à l'étude morphologique et palynologique de la Grande Dune du Pilat (Gironde). P. V. Société Linéenne, Bordeaux, 98, 12p. (1961).

[20] Sargos, R.: Contribution à l'histoire du boisement des Landes de Gascogne. 836 p. Delmas Edit., Bordeaux (1949).

[21] Ters, M.: Les lignes de rivage quaternaire de la côte atlantique. 333−341. La Préhist. Franç., 9me Congrés de l'UISPP., C.N.R.S. Edit., (1976).

Authors' addresses: Dr. J.-M. Froidefond, Chargé de Recherche C.N.R.S., and Prof. Dr. R. Prud'homme, URA 197, (C. N. R. S.) Département de Géologie et Océanographie, Université de Bordeaux, 1, Av. des Facultés, 33405 Talence cedex, France.

Acta Mechanica (1991) [Suppl] 2: 161—170
© by Springer-Verlag 1991

Controls on aeolian sand sheet formation exemplified by the Lower Triassic of Helgoland

L. B. Clemmensen, Copenhagen, Denmark

Summary. The basal part of the Middle Buntsandstein (Lower Triassic) on Helgoland in the southern North Sea is composed of lacustrine redbeds with thin white beds of aeolian sand sheet origin. The aeolian deposits formed during lake low-stands, and sand sheet formation was caused by a restricted sand supply perhaps in combination with frequent wind changes and high-speed wind events. With time aeolian sedimentation was progressively influenced by brief flood events or by minor lake transgressions. Ultimately the sand sheets were covered by fine-grained lake deposits during a new phase of lake high-stand. Preservation of the aeolian deposits records the overall low-energy of the littoral processes, and early cementation. The aeolian deposits define sedimentary cycles of various thickness, and it is suggested that the cyclicity records Early Triassic orbital climatic forcing.

1 Introduction

Much research in recent years has been directed towards an understanding of the complex interactions between aeolian and marine, or aeolian and fluvial realms (e.g. Chan and Kocurek [4], Langford [14], Langford and Chan [15]). Less attention so far has been paid to aeolian-lacustrine process interactions.

In this paper the nature of aeolian sand sheet deposits from a Lower Triassic desert-lake system (Helgoland, southern North Sea, Fig. 1) is described. The aeolian sand sheet and associated lake deposits are well exposed and allow a reconstruction of aeolian and littoral processes, sand sheet dynamics and lake-level changes.

2 Description of aeolian sand sheet deposits

The island of Helgoland is mainly composed of fine-grained shallow-water red-beds from the Middle Buntsandstein (Lower Triassic), and encompasses the Uppermost Volpriehausen-Folge, the Detfurth-Folge, the Hardegsen-Folge and the Solling-Folge (Binot and Röhling [2]).

The upper part of the Volpriehausen-Folge and the basal part of the overlying Detfurth-Folge contain several white and practically uncemented sand beds ("Katersande"). An aeolian origin for these well-sorted sand beds was already proposed by Clemmensen [5], and additional evidence is given here to support this conclusion. Due to the dominance of horizonal or low-angle wind-ripple stratification, the sediments in question seem to represent ancient aeolian sand sheet deposits (cf. Kocurek and Nielson [12]). The aeolian sediments are composed of fine-to medium-grained quartz sand. The coarser grains are rounded to well rounded and nearly always possess frosted surfaces. Grain-size analyses of typical

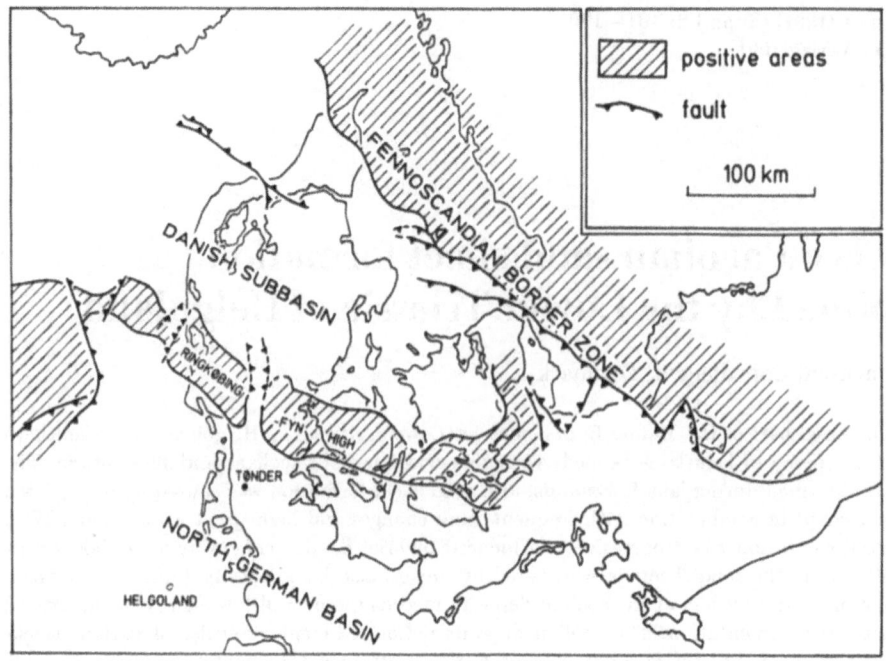

Fig. 1. Structural map of Denmark and nearby areas in Early Triassic (based on Berthelsen [1] and Fine [9]), and the location of the studied sediments at Helgoland

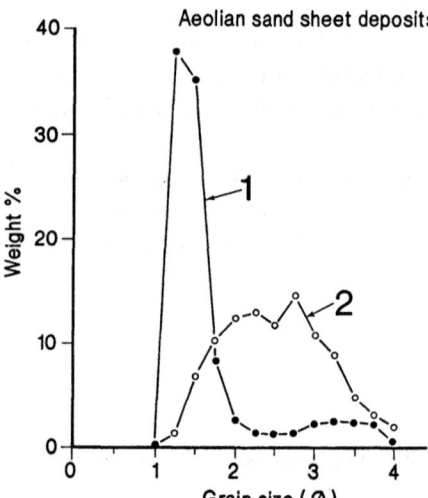

Fig. 2. Frequency curves of typical samples from the lower Triassic aeolian sand sheet deposits, Helgoland. 1 wind-ripple strata, 2 grainfall strata

samples (Table 1; Folk and Ward [10]) indicate that the mean grain-size varies from 1.47 ∅ (0.36 mm) to 2.38 ∅ (0.19 mm) with wind-ripple strata being significantly more coarse-grained than grainfall strata (Fig. 2). They also indicate that the sand is moderately well to well sorted (standard deviation between 0.67 and 0.37) and fine skewed to strongly fine skewed (skewness between +0.10 and +0.55). A comparison with modern aeolian deposits from the Kuwaiti desert (Khalaf [11]) shows that the ancient sand sheet deposits have a mean grain-size that compares well with that of modern active sand sheet deposits which have mean grainsizes between 0.45 and 2.51 ∅ with an average of 1.95 ∅. On the other hand the

Fig. 3. Simple aeolian sand sheet deposit. Note the sharp basal erosion surface (sand-drift surface) and the gradational upper contact to overlying lake deposits. Volpriehausen-Folge, Helgoland

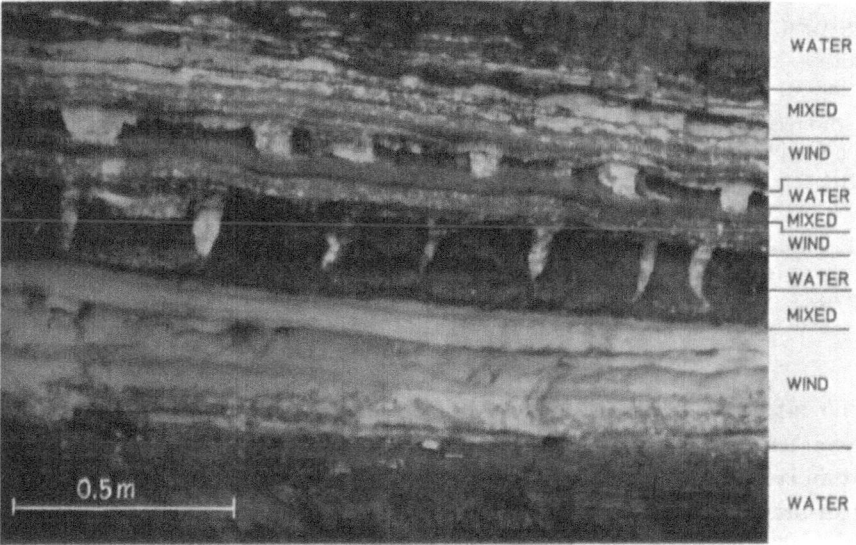

Fig. 4. Composite aeolian sand sheet deposit. The sand sheet is composed of three wetting-upward sub-cycles each initiated by wind-formed sand and topped by waterlaid deposits. Note the sand-filled desiccation cracks at the base of the middle and upper sub-cycles. Volpriehausen-Folge, Helgoland

Helgoland sand sheet deposits are markedly better sorted than the Kuwaiti modern counterparts having sorting values between 0.46 and 1.83 $\varnothing$ with an average of 1.06.

The sand sheet deposits are composed mainly of low-angle or horizontally stratified wind-ripple deposits (Figs. 3, 4), but higher angle cross-stratified aeolian deposits with grainfall and sandflow strata are locally common. The sand sheet beds are characterized by a large lateral variation in sedimentary structures, although the thickness of individual beds remains relatively constant. In some of the thin sand sheet deposits salt ridge structures (Fryberger et al. [10]) are common. In the upper part of most of the sand sheet deposits

Table 1. Grain-size parameters (standard sieving technique), aeolian sand
sheet deposits (L. Triassic), Helgoland

Sample	Mean	Standard deviation	Skewness
	∅	∅	∅
1	1.47	0.37	+0.37
2 Wind-ripple strata	1.47	0.54	+0.58
3	1.50	0.51	+0.71
4 Grainfall strata	2.38	0.67	+0.10

Table 2. Data on modern orbital periods (Olsen [14]) and cycle thicknesses in the Lower Triassic
lake deposits on Helgoland. Abbreviations of orbital periods: *P* precession of the equinoxes; *O* obliquity
cycle; *E*1, *E*2, and *E*3 eccentricity cycles. The precession cycle is actually composed of two periods
of 19.000 and 23.000 years; in the Early Triassic these periods were shortened to c. 17.800 and c.
11.300 years (A. Berger, pers. comm. 1991)

Item	Orbital periods				
	P	*O*	*E*1	*E*2	*E*3
Present-day periods ($\times 10^3$ years)	21.7	41.0	93.0	123.0	412.1
Average thicknesses of Helgoland cycles (m)	2.0	—	—	11.5	35.5
Ratio of modern precession period to other modern orbital periods	1.0	1.9	4.4	5.8	19.0
Ratio of thin Helgoland cycles to thicker cycles	1.0	—	—	5.8	17.8
Helgoland cycles, calculated depositional time ($\times 10^3$ years)	20.3	—	—	117.6	364.1

there is much evidence of water-reworking. Water-formed structures include small scours
tilled with reworked aeolian sand and intraformational clay pebbles, wave ripples, mud
drapes, and thin layers with fine-grained lacustrine red-beds.

The aeolian sand sheet deposits are underlain by well defined, relatively flat erosion
surfaces termed sand-drift surfaces by Clemmensen and Tirsgaard [7]. These sand-drift
surfaces are classified into two types: (1) main sand-drift surfaces which are laterally extensive and overlain by 0.20—1.0 m thick units of sand sheet deposits, and (2) minor sand-drift
surfaces which are less continuous and only overlain by 0.02—0.1 m thick aeolian or reworked aeolian deposits. The uppermost contact of main sand sheet deposits is typically
gradational and characterized by an up to 10—30 cm thick zone of intimately interbedded
aeolian and shallow-water deposits.

Based on careful studies of the vertical variation in lithology and stratification types,
it is possible to define simple and composite sand sheets. The simple sand sheets display a
simple 0.20—1.0 m thick wetting-upward sequence composed of basal dry aeolian sand and
overlying interbedded aeolian and water-laid deposits topped by shallow-water mud-rich
red-beds (Fig. 3). The composite sand sheets also possess an overall wetting-upward sequence, but contains two or three 0.20—0.40 m thick wetting-upward sub-cycles (Fig. 4).

Each sub-cycle has a basal unit with dry aeolian sand, a middle zone with interbedded aeolian and shallow-water deposits and an upper zone with shallow-water mud-rich red-beds. Typically the amount of dry aeolian sand in the sub-cycles decreases upwards. In places, however, much of the dry aeolian deposits has been scoured and reworked by water.

The aeolian sand sheets and associated sand-drift surfaces divide the lake-sequence into well-developed cycles (Clemmensen and Tirsgaard [7]). Main cycles initiated by a main sand-drift surface are spaced between 11.0 and 11.9 m with an average of 11.5 m. Minor cycles, initiated by a minor sand-drift surface, are spaced between 1.5 and 2.4 m with an average of 2.0 m. On a large scale the aeolian and related lake-shore sands define megacycles (Grosszyklen, Binot and Röhling [2]). Two of the better exposed megacycles (Grosszyklus II and III) have thicknesses of 33 and 38 m. Thus the lake deposits are clearly cyclic with cycles having average thicknesses of 2.0, 11.5 and 35.5 m. The ratios of the thicknesses of the cycles are 1.0:5.8:17.8 (Table 2).

3 Description of lake deposits

The lake deposits of the Volpriehausen- and Detfurth-Folgen are composed of moderately well cemented, red-coloured heteroliths and mudstones with grain-sizes varying from fine-grained sand to clay. The mudstones are commonly massive, whereas the heteroliths are characterized by wave-ripple formsets and associated cross-lamination. The wave-ripple structures often form lenticular or flaser bedding. Associated structures are horizontal and undulatory lamination. Current-ripple cross-lamination is rare. Desiccation cracks are abundant and halite or gypsum pseudomorphs are common in many beds. The desiccation cracks are typically concentrated in the upper part of the minor cycles (cf. Binot and Röhling [2]) suggesting a shallowing of the water-body prior to aeolian sedimentation. Ripple marks are common on exposed bedding-planes. The ripple-index of these ripples lie between 5 and 25, and their symmetry-index varies between 1 and 4.5 with the majority between 1 and 2 (Bruun-Petersen and Krumbein [3]; and author's observations). Based on these data Bruun-Petersen and Krumbein [3] conclude that nearly all ripples are wave-generated; a few ripples may be wind-formed. The majority of the wave ripples trend NE to SW, while a subordinate group trend WNW—ESE (Clemmensen [5]).

The lake deposits form the bulk of the discussed sequence and form units c. 11 m thick separated by well-developed aeolian sand sheet deposits.

4 Discussion

Depositional environment

The depositional environment of the Volpriehausen- and Detfurth-Folgen has been much debated (e.g. Wurster [7], Bruun-Petersen and Krumbein [3], Clemmensen [5], Binot and Röhling [2]). In agreement with the ideas of Clemmensen [5] it is suggested here that the sediments formed in a large desert-lake system (Figs. 5, 6). The lake fluctuated considerably in size and during lake low-stands large parts of the lake seem to have dried up completely. The lake system may occasionally have been influenced by marine water, but the longterm dynamics of the lake seem to have been controlled by internal drainage.

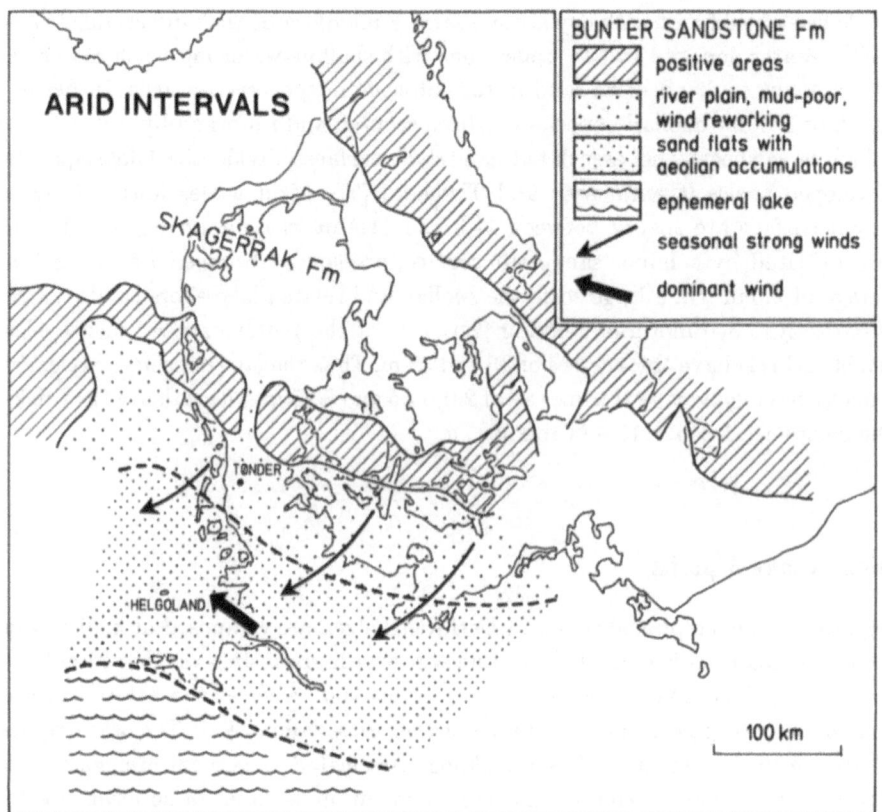

Fig. 5. Tentative palaeogeographical reconstruction of the northern part of the North German Basin during deposition of the Volpriehausen-Folge and early Detfurth-Folge. The figure portrays a situation with maximum extension of the dry sand flats during arid climatic intervals

Sand sheet formation

The white sandstones ("Katersande") have been interpreted as fluvial (e.g. Wurster [15]) and aeolian (Clemmensen [5]). The present data on grain texture, grain-size characteristics and in particular sedimentary structures strongly support an aeolian origin. The dominance of horizontal or low-angle wind-ripple stratification classifies the aeolian environment as sand sheets. These ancient sand sheets, however, were not completely flat but were characterized by scattered protodunes (low-angle grainfall and wind-ripple stratification) and rare barchan dunes (high-angle sandflow stratification).

Based on a study of six modern aeolian sand sheet deposits Kocurek and Nielson [12] concluded that factors favourable for sand sheet development are: (1) a high water table, (2) surface cementation, (3) periodic flooding, (4) a coarse grain size, and (5) vegetation. They also suggest that vegetation and a coarse grain size (a mode greater that 0.65 mm) are the prime factors causing sand sheet formation.

The Triassic sand sheet deposits on Helgoland do not contain any direct evidence of ancient vegetation, and it is visualised that the sand sheets were at best sparsely vegetated. The mean grain size of the ancient aeolian deposits lies between 0.19 and 0.36 mm (2.38 to 1.47 $\varnothing$). Grains of this size are easily moved by moderate to strong winds and many modern and ancient dune deposits are equally coarse. Apparently, factors other than vegetation and grain size caused sand sheet rather than dune formation on Helgoland. It is suggested

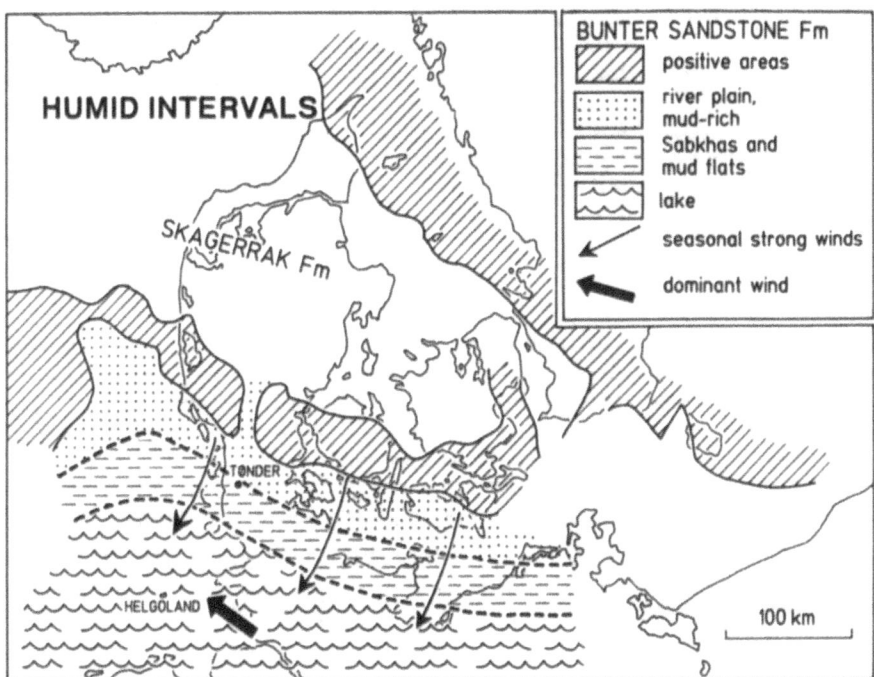

Fig. 6. Tentative paleogeographical reconstruction of the northern part of the North German Basin during deposition of the Volpriehausen-Folge and early Detfurth-Folge. The figure portrays a situation with maximum extension of the lake during humid climatic intervals. The climatic fluctuations were periodic and probably controlled by variations in the earth's orbit (Milankowitch cycles)

that a lack of adequate sand supply was the main control on aeolian sedimentation. The lake deposits contain very few medium-sized sand grains and could not have been the primary source of the aeolian deposits. A more likely source is fluvial and sand flat deposits situated to the north of Helgoland (Fig. 5; cf. Clemmensen [6]). Transport distance (tens of km) in combination with a relatively high water-table and/or early surface cementation in the area of deposition would seriously restrict the amount of dry and loose sand available for aeolian processes on Helgoland. According to Kocurek et al. [14] the level of sand-saturation is fundamental in controlling aeolian sedimentation. Having a limited sand supply and assuming frequent high-energy wind events on the Helgoland sand sheets, the overall level of sand-saturation would be low and dune forms were not likely to develop. Frequent wind shifts would also inhibit the formation of barchan dunes. In addition, periodic flooding especially during the final phases of aeolian sedimentation may have washed away any existing dunes and also promoted sand sheet formation.

Thus it is envisaged that the aeolian sand sheets formed on lake sediments that were exposed due to a fall in the lake level. Initial wind erosion of the lake sediments was followed by a period of aeolian sand drift but, due to a limited supply of dry and loose sand, dune building was rarely achieved and most sand was deposited as aeolian sand sheets.

Lake wave climate and water depth

Estimates of palaeowave conditions can be obtained from the grain size and spacing of wave ripple marks (Diem [8]). In calculating the ancient wave conditions only orbital ripples can be used. In the present study therefore, only ripples with a ripple-index less than

Table 3. Estimation of palaeowave climate and palaeodepth, lake deposits (L. Triassic), Helgoland

Ripple length cm	Grain size mm	do cm/s	Ut cm/s	Um cm/s	T min s	T max s	Lt_∞ m	L max m	L min m	h max m	H max m
3.0	0.1	4.6	8.7	22.7	0.6	1.6	4.2	4.2	0.3	2.2	0.6
6.2	0.1	9.5	10.4	32.6	0.9	2.9	12.9	12.9	0.7	7.5	1.8

For exploration of abbreviations see Table 4

Table 4. Analytical expressions for palaeodepth and palaeowave climate reconstructions from wave ripple marks (Diem [8]). λ = wave ripple spacing, D = mean grain size, p = density of water, p_s = density of sediment, g = acceleration due to gravity, v = kinematic viscosity

Wave characteristics	Symbol	Expression
Bottom orbital diameter	do	$do = \lambda/0.65$ provided $\lambda < 0.0028 \cdot D^{1.68}$
Threshold velocity	U_t	$U_t^2 = 0.21(do/D)^{1/2}(p_s - p)\,gD/p$, $D < 0.5$ mm
Bottom orbital velocity (maximum)	Um	$U_t < Um \leq \sqrt{0.112g\,do}$
Wave period	T	$\pi\sqrt{8.9do/g} \leq T < \pi do/U_t$
Maximum possible deep water wave length	$L_{t\infty}$	$L_{t\infty} = (\pi g do^2)/(2U_t^2)$
Wave length	L	$L_{\mathrm{max,min}} = \left(L_{t\infty}/\sqrt{2}\right)\sqrt{1 \pm \sqrt{1 - 80.4 \cdot U_t^4/(g^2 do^2)}}$
Water depth	h	$h < (L_{\mathrm{max}}/2\pi)\,\mathrm{arcosh}\,(0.142 L_{\mathrm{max}}/do)$
Wave height	H	$H < do\,\sinh(2\pi h_{\mathrm{max}}/L_{\mathrm{max}})$

8.0 were taken. These wave ripples have wavelengths between 3.0 and 6.2 cm and mean grain sizes around 0.1 mm (Table 3).

Using the analytical expressions of Diem [8] shown in Table 4, it is possible to estimate palaeowave climate and palaeodepth in the Triassic lake (Table 3). Calculated wave periods vary between 0.6 and 2.9 sec and this classifies the lake as being of low to moderate energy. Calculated values of maximum water depth lie between 2.2 and 7.5 m (Table 3). The close interbedding of wave ripple structures and desiccation cracks, however, suggest that actual maximum water depth was considerably less and perhaps rarely more than 1 m. Thus this ancient lake was fairly shallow and of low energy even when it had its maximum development. Episodic storms may have occurred as suggested by the presence of horizontal and undulatory (hummocky) lamination.

Sand sheet dynamics and lake level changes

The regular alternation of aeolian sand sheet and lake deposits suggests repeated fluctuations in lake level (Figs. 5, 6). The cyclic sequence can not be explained by tectonism (cf. Binot and Röhling [2]). During lake low-stands the previous lake deposits on Helgoland became exposed and subject to aeolian processes. Initial deflation (probably only minor erosion of lake sediments) was followed by aeolian sand transport and sand sheet formation, but the exact time relations between initial erosion and aeolian deposition is difficult to estimate.

Early aeolian deposition took place during relatively dry surface conditions but with time aeolian deposition took place during increasingly wet conditions and finally aeolian deposition was repeatedly interrupted by water-reworking. Due to the gradual contact between aeolian and overlying lacustrine deposits, it is suggested here that reworking of the aeolian deposits actually took place during brief storm events causing wash-over phenomena at the lake shore aeolian sand flats. Ultimately the aeolian sand sheets were transgressed by the lake and preserved beneath fine-grained lake deposits. The preservation of the aeolian deposits is explained by the overall low energy of the littoral processes and by early cementation of the sand sheet deposits. Any signs of this early cement (probably calcite and gypsum; cf. Fine [9]), have, however, been leached out by ground water.

The composite nature of many aeolian sand sheet deposits actually indicates a stepwise nature for the lake transgressions.

The well-developed cyclic pattern of the aeolian and associated lake deposit indicates systematic lake level fluctuations of various periodicity. The calculated ratios between thicknesses of the observed cycles (Table 2) and the great lateral continuity of the cycles (cf. Binot and Röhling [2]) suggest that these fluctuations in lake level were the result of orbital climatic forcing (Milankovitch cycles).

Based on the knowledge that Buntsandstein (total thickness of 975 m) on Helgoland was deposited in 10 m.y., the depositional time of each cycle can be calculated assuming a constant subsidence rate and sedimentation rate (Table 2). The resulting values are good support for the climatic interpretation of the cycles. Triassic lacustrine cycles of the Newark Supergroup (eastern North America) have also been interpreted as the result of orbital climatic forcing (Olsen [17]). The latter cycles, however, do not contain any aeolian facies.

5 Conclusions

1. The Middle Buntsandstein on Helgoland contains regularly spaced aeolian sand sheet deposits. Aeolian sedimentation probably reflects a limited supply of dry and loose sand.

2. The aeolian sand sheets were formed during lake lowstands, became partly reworked during rising lake levels, and ultimately preserved beneath lake sediments during lake highstands.

3. Fluctuations in lake level were probably linked to Early Triassic orbital climatic forcing which controlled the amount of precipitation in the region.

Acknowledgements

I would like to thank J. T. Møller and K. Pye for critically reading the manuscript and J. Bailey for correcting the English of the article. G. Pedersen and D. Pugliese typed the manuscript, and R. Madsen drafted the figures.

References

[1] Berthelsen, F.: Lithostratigraphy and depositional history of the Danish Triassic. Danm. Geol. Unders. (Ser. B) 4, 59 pp. (1980).
[2] Binot, F., Röhling, H.-G.: Lithostratigrafie und natürliche Gammastrahlung des Mittleren Buntsandsteins von Helgoland. — Ein Vergleich mit der Nordseebohrung J/18-1. 2. Dt. Geol. Ges. 139, 33—49 (1988).

[3] Bruun-Petersen, J., Krumbein, W. E.: Rippelmarken, Trockenrisse und andere Seichtwasser-merkmale im Buntsandstein von Helgoland. Geol. Rundschau 64, 126—143 (1975).

[4] Chan, M. A., Kocurek, G.: Complexities in eolian and marine interactions: processes and eustatic controls on erg development. Sedimentary Geol. 56, 283—300 (1988).

[5] Clemmensen, L. B.: Triassic lacustrine red-beds and palaeoclimate: the "Buntsandstein" of Helgoland and the Malmros Klint Member of East Greenland. Geol. Rundschau 68, 748—774 (1979).

[6] Clemmensen, L. B.: Desert sand plain and sabkha deposits from the Bunter Sandstone Formation (L. Triassic) at the northern margin of the German Basin. Geol. Rundschau 74/3, 519—536 (1985).

[7] Clemmensen, L. B., Tirsgaard, H.: Sand-drift surfaces: A neglected type of bounding surface. Geology 18, 1142—1145, (1990).

[8] Diem, B.: Analytical method for estimating palaeowave climate and water depth from wave ripple marks. Sedimentology 32, 705—720 (1985).

[9] Fine, S.: The diagenesis of the Lower Triassic Bunter Sandstone Formation, onshore Denmark. Danm. Geol. Unders. (Ser. A) 15, 51 pp. (1986).

[10] Folk, R. L., Ward, C.: Brazos River Bar: a study in the significance of grain size parameters. J. Sedim. Petrol. 27, 3—26 (1957).

[11] Fryberger, S. G., Al-Sari, A. M., Clisham, T. J., Rizvi, S. A. R., Al-Kinai, K. G.: Wind sedimen-tation in the Jafurah sand sea, Saudi Arabia. Sedimentology 31, 413—431 (1984).

[12] Khalaf, F.: Textural characteristics and genesis of the eolian sediments ih the Kuwaiti desert. Sedimentology 36, 253—271 (1989).

[13] Kocurek, G., Nielson, J.: Conditions favourable for the formation of warm-climate aeolian sand sheets. Sedimentology 33, 795—816 (1986).

[14] Kocurek, G., Townsley, M., Yeh, M., Sweet, M., Havholm, K.: Dune and dune-field development stages on Padre Island, Texas: effects of lee airflow and sand-saturation levels and implications for interdune deposition. (1990) Extended abstracts. Nato advanced research workshop on sand, dust and soil in their relation to aeolian and littoral processes.

[15] Langford, R. P.: Fluvial-aeolian interactions. Part I. Modern systems. Sedimentology 36, 1023 — 1035 (1989).

[16] Langford, R. P., Chan, M. A.: Fluvial-aeolian interactions. Part II. Ancient systems. Sedimento-logy 36, 1037—1051 (1989).

[17] Olsen, P.: A 40-million-year lake record of Early mesozoic orbital climatic forcing. Science 234, 842—848 (1986).

[18] Wurster, P.: Kreuzschichtung in Buntsandstein von Helgoland. Mitt. Geol. Staatsinst. Hamburg 29, 61—65 (1960).

Author's address: Dr. L. B. Clemmensen, Institute of General Geology, Øster Voldgade 10, 1350 Copenhagen K, Denmark.

Acta Mechanica (1991) [Suppl] 2: 171—181

Beach deflation and backshore dune formation following erosion under storm surge conditions: an example from Northwest England

K. Pye, Reading, United Kingdom

Summary. Field observations following a major storm surge on 26 February 1990 on the Sefton Coast in Northwest England showed that the rate of aeolian sand transport and backshore dune construction varied significantly alongshore in response to variations in beach morphology and sand wetness. These variations, in turn, were determined by longshore variations in the character of marine sediment transport processes and the distribution of subsurface silt and peat formations which outcrop locally on the foreshore. The grain size of the beach sands was found to be very uniform and was not a factor influencing spatial variations in sand transport rate. Since the beach sediments consisted of well-sorted and very well-sorted fine sands, partly reworked from the dunes during the storm surge, deflation was not grain-size selective during the period of post-surge dune recovery.

1 Introduction

Coastal dunes act as natural sea defences which absorb wave energy during storms. On the Sefton Coast in Northwest England (Fig. 1), a belt of coastal dunes up to 2 km wide protects more than 150 km^2 of high grade agricultural land from marine flooding [1], [2]. Following a period of rapid coastal progradation all along the coast in the second half of the last century, the dune frontage at Formby Point has been eroding since about 1906 (Fig. 2). Most of the dune erosion occurs when strong winds and destructive waves coincide with high spring tides, usually during the period September—March [3]. Very damaging storm tides, which often cause more than 5 m of dune recession, occur on average every 5 or 6 years. The rate of recovery of the frontal dunes following such a storm has generally been insufficient to prevent a long-term net erosional trend between Lifeboat Road and Fisherman's Path (Fig. 2). However, further north, between Ainsdale and Southport, and on the south side of Formby Point between Alexandra Road and the mouth of the River Alt, erosion during storms is generally less severe and the rate of dune recovery has been sufficiently rapid to maintain a net progradation trend throughout this century [1], [4], [5]. In view of increasing concern about the possibility that dune erosion at Formby may spread or accelerate in response to sea level rise or an increase in storminess associated with greenhouse warming, studies are in progress to achieve a better understanding of beach-dune interaction in this area. This paper summarises the results of observations made following a recent severe erosion event in February 1990.

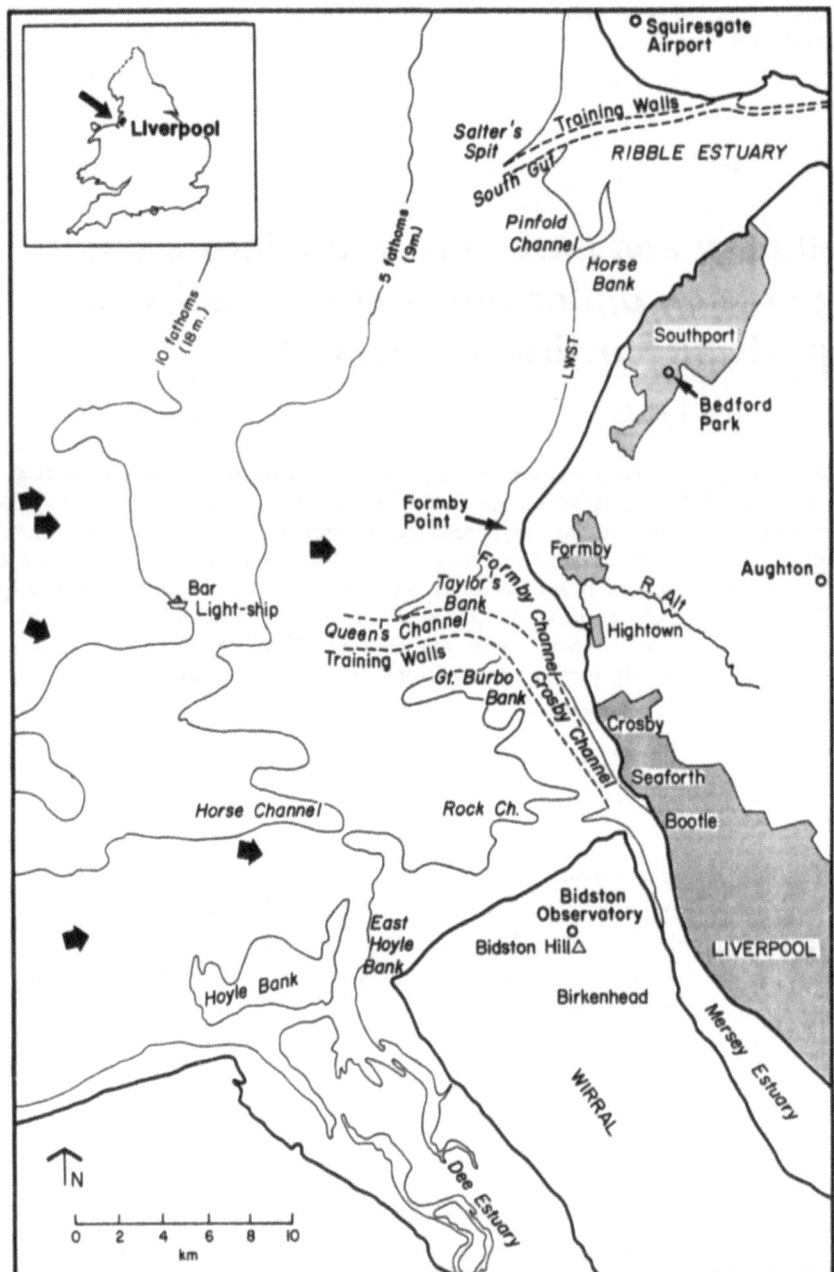

Fig. 1. Location of the Sefton Coast in northwest England. The large arrows indicate the direction of sea bed sediment transport indicated by the orientation of sand waves

2 Dune erosion during the storm surge of 26 February 1990

The storm surge of 26 February 1990 was associated with the passage of a vigorous depression across northern Scotland and the North Sea during the period 25—27 February (Fig. 3). From 0000—0600 hrs on 26 February the mean hourly wind speed at Squiresgate Airport, Blackpool (Fig. 1) increased from 18 knots (33 km h^{-1}) to 45 knots (73 km h^{-1}), while the

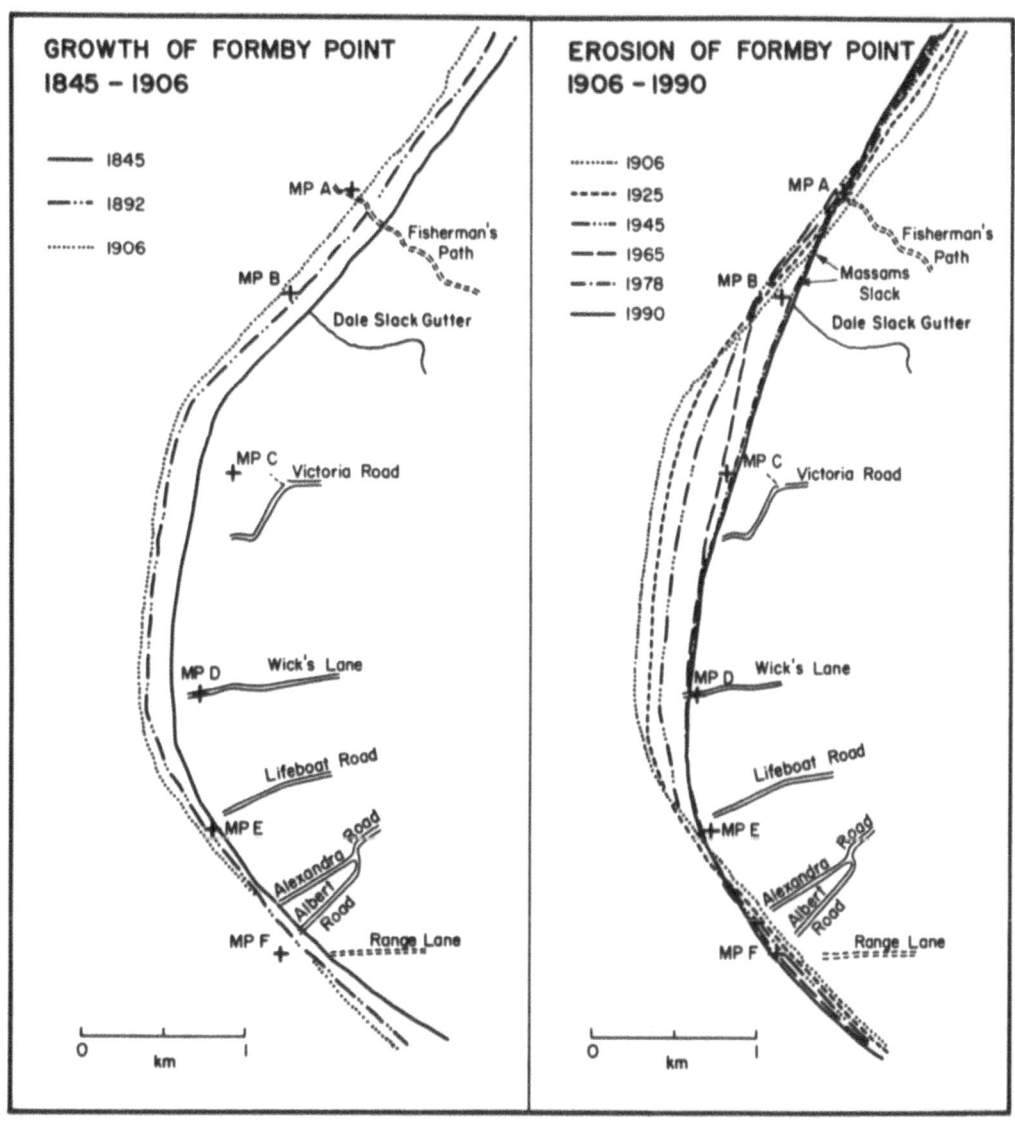

Fig. 2. Plan diagrams showing the growth of Formby Point 1845–1906 and subsequent erosion in the period 1906–1990. Based on Ordnance Survey maps, air photographs, and field measurements taken by Sefton Borough Engineer & Surveyor's Department

direction shifted from southwesterly to westerly (Fig. 4). These strong onshore winds produced a surge which raised the height of predicted high water at 1200 hrs. by approximately 1 m along the coast between Morecambe Bay and North Wales. The resultant high tide exceeded 6.3 m O.D. (Ordnance Datum) at Liverpool, 6.4 m O.D. at Southport and 6.4 m O.D. at Heysham. Structural damage was caused along the promenade at Southport and Crosby, and the sea defences were seriously breached at Towyn in North Wales, causing flooding to several thousand homes.

The dune frontage between Hightown and Southport suffered severe erosion, amounting to 6.0 m at Albert Road, 8.3 m at Lifeboat Road, 11.1 m at Wick's Lane, 13.6 m a Victoria Road, 7.5 m at Fisherman's path and 6.0 m between Ainsdale and Southport. Near

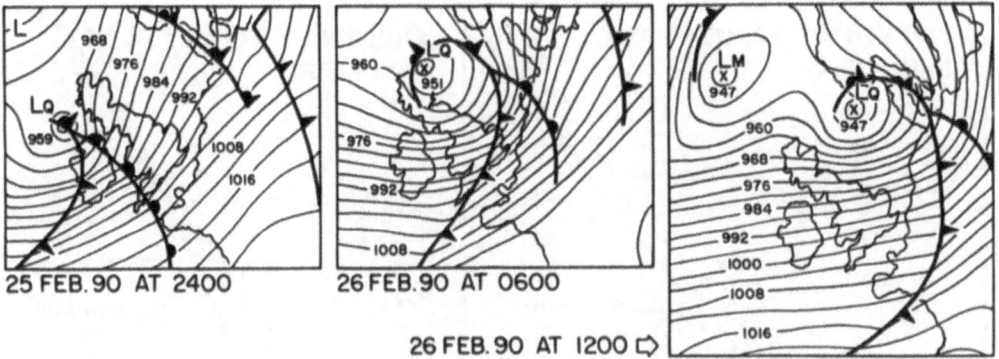

Fig. 3. Synoptic charts showing the path taken by the depression responsible for the storm surge on 26th February 1990 (based on the Daily Weather Reports of the U.K. Meteorological Office)

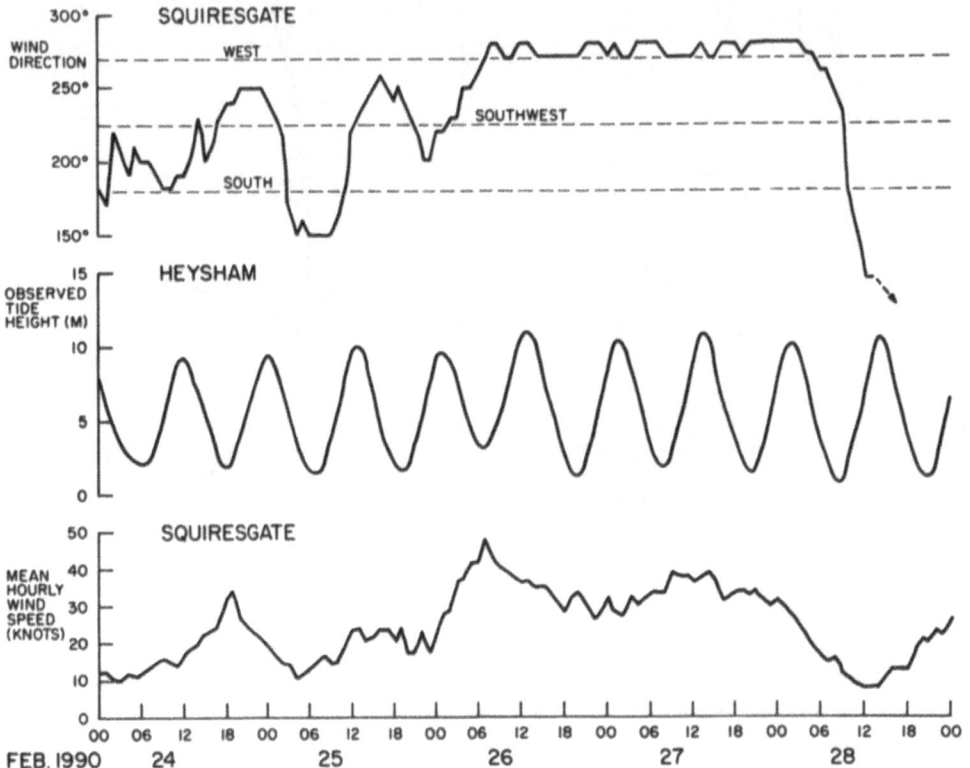

Fig. 4. Variations in wind direction recorded at Squiresgate Airport, Blackpool (top), mean hourly wind speed (bottom), and tidal height (relative to Chart Datum) recorded at Heysham (middle) during the period 24th—28th February 1990. Wind data supplied by the Meteorological Office. Tide data supplied by the Institute of Oceanographic Sciences Proudman Laboratory

Fisherman's Path, erosion formed dune cliffs 8—10 m high (Figs. 5, 6, and 7). At Massam's Slack, where previous erosion has truncated the ends of dune ridges created artificially during the 1920's, waves overtopped the frontal ridge and flooded the slack behind.

Strong westerly winds continued during the period 27—28 February (Fig. 4), although a slight reduction in velocity and a reduction in wave height meant that the very severe erosion experienced on the morning of the 26 February was not repeated.

Fig. 5. Dune cliff eroded during the storm of 26th February near Fisherman's Path. Note subsequent cliff degradation is taking place as a result of dry sandflows.

Fig. 6. Blocks of dune sand bound by vegetation littering the upper beach north of Fisherman's Path after the February storm

Beach deflation and backshore dune formation following the storm surge

The high storm tides created a platform of wave-reworked sand to seaward of the eroded dune face around Formby Point (Fig. 7). Extensive saltaton was observed on this surface during low tides on the 26th, 27th and 28th February (Fig. 8). However, much of the blown sand was removed by the following high tides on these days, and it was not until the winds and tides began to drop after 1st March that significant amounts of sand accumulated in front of the eroded dune line. The period 1st—7th March was one of moderate, predominantly onshore southwesterly winds which deflated the increasing width of beach being left uncovered by the falling high tides.

Fig. 7. Wave eroded sand platform north of Lifeboat Road. An outcrop of organic-rich dune slack peat is exposed along the upper beach

Fig. 8. Active saltation on the shore near Fisherman's Path

The amount of sand which accumulated during this period was observed to vary significantly along the shore. Between Lifeboat Road and Wick's Lane, the amount of deflation and backshore sand accumulation was limited by an outcrop of organic-rich dune slack peat exposed over much of the upper foreshore (Fig. 7). Between Dale Slack Gutter and Fisherman's Path deflation was limited by a high beach water table which outcrops close to the dune line due to the presence of a sub-surface layer of Holocene sandy silts. South of Lifeboat Road, where the foreshore was divided into a 30 m-wide upper section and a much wider (> 1 km), generally wet, lower section with well-developed ridge and runnel topography, deflation was restricted to the lower part of the upper foreshore, where numerous scour-remnant ridges were formed (Fig. 9). The height of these features indicated a maxit mum amount of vertical lowering by deflation of 3—4 cm in the period 26 February—1s-

Fig. 9. Scour-remnant ridges on the deflated part of the upper foreshore near Alexandra Road

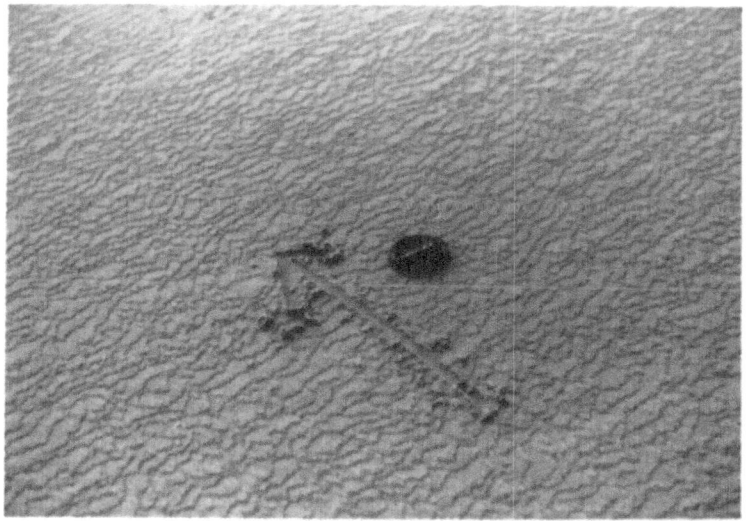

Fig. 10. Adhesion ripples formed by trapping of blown sand on the surface of a wet runnel, off Albert Road

March. On the lower foreshore, localised deflation occurred on the ridge crests, but most of the sand blown from these sources was trapped in the intervening runnels, forming adhesion ripples (Fig. 10). The backshore sand accumulation in this area was typically 5 m wide and 0.3—0.5 m thick, tapering seawards (Fig. 11). The wind in this area approached the shore almost normally, heaping the backshore sand into a series of coalesced eeho dunes with barchanoid wings, many of which displayed climbing behaviour.

A similar pattern of deflation and backshore sand accumulation was observed between Victoria Road and Dale Slack Gutter, where the beach morphology was broadly similar to that south of Lifeboat Road. However, in this area the wind blew at a high oblique angle across the beach and formed a series of seif-like dunelets with crests aligned at an angle of 40—45° to the eroded dune cliff-line (Fig. 12).

Fig. 11. Backshore blown sand accumulation between Albert Road and Range Lane

Fig. 12. Seif-like dunelets on the backshore north of Victoria Road

Grain-size variations

Samples of sand were collected along 15 shore-normal transects between Range Lane and Southport Pier (Fig. 13) in order to establish (a) whether there were any significant long-shore variations in grain size, and (b) whether any such variations might be responsible for differences in the rate of aeolian sand transport and backshore dune formation. The sand samples were collected from a depth interval of 0—5 cm using a trowel. Along each transect samples were collected from the following locations: (a) the mid part of the lower foreshore (often wet, with little evidence of deflation); (b) the upper part of the lower foreshore (in many places wet with evidence of adhesion ripple development); (c) the lower part of the upper foreshore (often areas of deflation with well-developed scour-remnant ridges); and (d) the crest of backshore dunes. All samples were washed repeatedly in the laboratory to

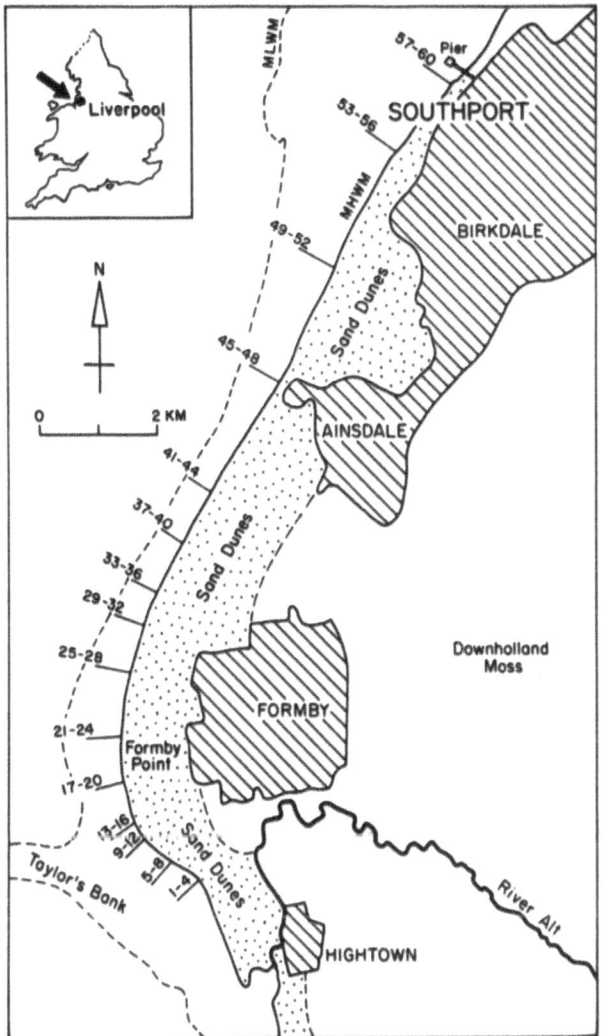

Fig. 13. Location of the transects along which sediments were collected for grain size analysis

remove soluble salts, oven-dried at 105 °C, and sieved at quarter-phi intervals for 15 minutes using an Endecotts mechanical shaker and a nest of 15 half-height sieves. The mass-frequency data were subsequently processed using the PC grain size package Granny [6] which calculates both moment and graphical statistical parameters. A summary of the results is presented in Table 1 and a bivariate plot of mean size against sorting for all samples shown in Fig. 14.

All of the beach and dune samples can be classified as well or very-well sorted fine sands. No significant differences in mean size or sorting were found alongshore or between beach and backshore dune samples. Only in the extreme north, around Southport Pier, was the sorting found to be moderate, due principally to the presence of a higher proportion of comminuted shell debris in these sands. Most of the beach and backshore dune samples showed near-symmetrical distributions of slight positive skewness, although the more shelly samples from the Southport Pier and Range Lane areas possessed a small tail of coarse (mainly shell) particles and were consequently found to be slightly negatively skewed.

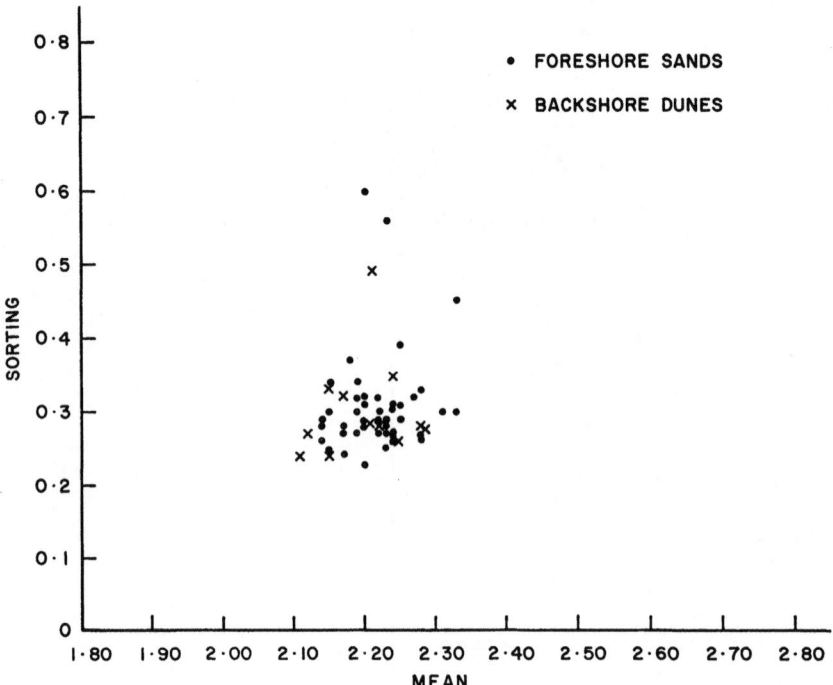

Fig. 14. Bivariate scattergram plot of mean (phi) grain size versus phi sorting (first and second moment parameters)

Table 1. Summary of grain size parameters for beach and backshore dune sand samples (n in each group $= 15$)

	Group A mid-lower foreshore (limited deflation)	Group B upper lower foreshore (adhesion ripples)	Group C lower upper foreshore (deflation area)	Group D backshore dune crests
Mean size (phi) (1st moment)				
average	2.22	2.21	2.21	2.21
range	2.14 to 2.34	2.19 to 2.23	2.15 to 2.25	2.11 to 2.31
Sorting (2nd moment)				
average	0.32	0.32	0.30	0.30
range	0.28 to 0.45	0.23 to 0.56	0.24 to 0.60	0.24 to 0.49
Skewness (graphic)				
average	+0.07	+0.03	+0.04	+0.06
range	−0.18 to +0.17	−0.27 to +0.13	−0.29 to +0.15	−0.41 to +0.30

The similarity between the grain size characteristics of the beach and backshore dune sands observed in this study is consistent with the results of previous work in coastal Brazil [7] and at Ainsdale [8] which indicated that aeolian transport in not grain-size selective when the source beach sediments are fine-grained and well-sorted.

Conclusions

Longshore variations in beach sand size were clearly not a significant factor influencing rates of deflation and backshore dune formation following the February 1990 erosional event on the Sefton coast. However, variations in beach morphology and sand wetness were controls of major importance. These, in turn, were governed by marine processes (waves, wave-generated longshore currents and tidal currents), and by the distribution of sub-surface silt and peat layers which determine the pattern of beach drainage.

Under the present sediment transport regime, the amount of sand moved onshore to Formby Point by constructive waves is inadequate to replace that lost offshore and alongshore during major storm events. To the north and south of Formby Point a net positive sediment budget is maintained partly by longshore transfer of sediment from the central section of coast between Lifeboat Road and just north of Fisherman's Path. Work is currently being undertaken to quantify the inputs and outputs of sand on different sectors of the coast.

Acknowledgements

Financial support provided by the Nuffield Foundation, NATO and the University of Reading Research Board is gratefully acknowledged. Reading University PRIS Contribution No. 065.

References

[1] Pye, K., Smith, A. J.: Beach and dune erosion and accretion on the Sefton Coast, Northwest England. J. Coastal Res. Spec. Issue 3, 33—36 (1988).
[2] Pye, K.: Physical and human influences on coastal dune development between the Ribble and Mersey estuaries, Northwest England. In: Coastal dunes: form and process. (Nordstrom, K. F., Psuty, N. P., Carter, R. W. G., eds.) Chichester: Wiley, pp. 339—359 (1990).
[3] Parker, W. R.: Sediment mobility and erosion on a multibarred foreshore, (Southwest Lancashire, U. K.). In: Nearshore sediment dynamics and sedimentation. (Hails, J. R., Carr, A. P., eds.) London: Wiley, pp. 151—177 (1975).
[4] Gresswell, R. K.: The geomorphology of the southwest Lancashire coastline. Geogr. J. 90, 335—349 (1937).
[5] Gresswell, R. K.: Sandy shores in South Lancashire. Liverpool: Liverpool University Press, (1953).
[6] Pye, K.: GRANNY — a package for processing grain size and shape data. Terra Nova 2, 588—590 (1990).
[7] Bigarella, J. J., Alessi, A. H., Becker, R. D., Duarte, G. M.: Textural characteristics of the coastal dune, sand ridge and beach sediments. Boletim Paran. Geocien. 27, 15—89 (1969).
[8] Vincent, P.: Differentiation of modern beach and coastal dune sands — a logistic regression approach using the parameters of the hyperbolic function. Sediment. Geol. 49, 167—176 (1986).

Author's address: Dr. K. Pye, Postgraduate Research Institute for Sedimentology, University of Reading, P.O. Box 227, Whiteknights, Reading RG6 2AB, United Kingdom.

Aeolian Grain Transport

Volume 1: Mechanics

Edited by O. E. Barndorff-Nielsen and B. B. Willetts

(Acta Mechanica/Supplementum 1)

1991. 79 figures. IX, 181 pages.
Soft cover DM 220,–, öS 1540,–
Reduced price for subscribers to "Acta Mechanica":
Soft cover DM 198,–, öS 1386,–
ISBN 3-211-82269-0

Prices are subject to change without notice

R. S. Anderson, M. Sørensen and B. B. Willetts:
A review of recent progress in our understanding of aeolian sediment transport

Saltation layer modelling:

R. S. Anderson and P. K. Haff: Wind modification and bed response during saltation of sand in air
I. K. McEwan and B. B. Willetts: Numerical model of the saltation cloud
M. Sørensen: An analytic model of wind-blown sand transport
M. R. Raupach: Saltation layers, vegetation canopies and roughness lengths

Observations of grain mobility:

G. R. Butterfield: Grain transport rates in steady and unsteady turbulent airflows
B. B. Willetts, I. K. McEwan and M. A. Rice: Initiation of motion of quarts sand grains
K. R. Rasmussen and H. E. Mikkelsen: Wind tunnel observations of aeolian transport rates
B. R. White and H. Mounla: An experimental study of Froude number of wind-tunnel saltation
M. A. Rice: Grain shape effects on aeolian sediment transport

Granular avalanche flow:

K. Hutter: Two- and three dimensional evolution of granular avalanche flow – Theory and experiments revisited

Springer-Verlag Wien New York